Inventory of potato variety collections in EC countries

Inventory of potato variety collections in EC countries

H.W.Kehoe
Agricultural Institute, Oak Park Research Centre, Carlow, Ireland

Published for the Commission of the European Communities by

A.A.Balkema / Rotterdam / Boston / 1986

CIP-DATA KONINKLIJKE BIBLIOTHEEK, DEN HAAG

Kehoe, H.W.

Inventory of potato variety collections in EC countries / H.W.Kehoe. – Rotterdam [etc.] : Balkema.
– Ill.
Publ. for the Commission of the European Communities. – With index.
ISBN 90-6191-633-X bound
SISO 632.4 UDC 633.49(4-672EEG) (083.8)
Subject heading: potatoes; EC.

EUR 10049

ISBN 90 6191 633 X

Published by A.A.Balkema, P.O.Box 1675, 3000 BR Rotterdam, Netherlands
Distributed in USA & Canada by: A.A.Balkema, P.O.Box 230, Accord, MA 02018

Table of contents

Preface

In 1978 the Standing Committee on Agricultural Research (SCAR), which is an advisory committee of the Commission of the European Communities, set up a Programme Committee on Plant Resistance and Use of Gene Banks as part of the second programme (1979–1983) on agricultural research. The Programme Committee in turn set up several working groups of which the Potato Working Group was one. The Potato Working Group decided to concentrate its work on two main tasks.

The first was the preparation of a list of descriptions of characteristics of cultivated potato varieties. The Programme Committee decided that this work was to be carried out in close cooperation with the International Board for Plant Genetic Resources (IBPGR) in Rome, as indeed was the case with the other crops dealt with in the programme.

The second task was the compilation of an inventory of all potato varieties in collections in the 10 Member States of the European Communities, a task which was undertaken with great enthusiasm by Dr. Henry Kehoe of An Foras Taluntais, Oak Park Research Centre, Carlow, Ireland, who assembled all the available information from colleagues in the Member States. The first draft was circulated to scientists in 1983.

For the third programme on agricultural research (1984–1988), a Programme Committee on Plant Productivity has been set up by the SCAR. This committee decided to complete the work in progress of the Potato Working Group.

The inventory was updated in 1985 and includes information on more than 1600 varieties in collections in the 10 Member States. In future this inventory will serve as a basis for work in compiling data on yield, quality and disease resistance characteristics of all varieties listed.

To reduce the workload and costs involved in the upkeep of genetic resources in the collections, I would suggest that the collections be screened for duplication, and evaluated by the Potato Working Group, and that cheaper storage facilities be found for this material.

Finally, this work has been done for the benefit of private and institutional breeders of this most important food and industrial crop, and will be of value not only within the Community but indeed, all over the world.

Ir. H.H.van den Borg
Co-ordinator EC Plant Productivity Group and former Co-ordinator of EC Plant Resistance and Use of Gene Banks Programme Committee 1979–1983

Acknowledgements

This inventory was compiled as part of the general programme on 'Plant Resistance and Use of Gene Banks' and was partly funded by the EC Commission. The author would like to acknowledge the co-operation and assistance of the members of the EC Potato Working Group of the Plant Resistance and Use of Gene Banks Committee who supplied details of potato variety collections in their own countries and made the necessary alterations to the draft inventory.

I would like to thank Ir. H.H.van der Borg, Co-ordinator of the main committee and Mr. P.J.O'Hare, Assistant Director, Agricultural Institute for their help and guidance.

I would especially like to thank the technical staff of the Agricultural Institute, Oak Park Research Centre, Carlow: Joan Dillon, Olivia Murphy, Eithne Waldron and Henry Walsh for their very considerable work in compiling the inventory. I would also like to thank Therese Coughlan and Catherine Doran for the typing of same.

Henry W.Kehoe
Co-ordinator Potato Working Group
EC Plant Resistance and Use of Gene Banks Programme Committee

Introduction and interpretation key

This inventory gives details of collections of potato varieties in EC countries and shows the health status of each variety. The varieties are compiled in alphabetical order. In addition details of the location of blight and potato cyst eelworm differential sets are given.

Interpretation key

The following codes were used in compiling the inventory
A = Variety present in tuber form and healthy (virus free).
B = Variety present in tuber form health status uncertain or diseased.
C = Variety stored *in vitro* and healthy.
D = Variety stored *in vitro* health status uncertain.
– = Variety not present in collection.

Location of collections

The Institute or Government Agency responsible for the various collections which may be contacted to obtain stocks of a variety in each country is as follows:

Belgium

> Ministère de l'Agriculture
> Station de Haute Belgique
> Rue de Serpont 48
> 6600 Libramont
> Chevingy

Denmark

1. Danish Potato Breeding Foundation
   The Breeding Station
   Grindstedvej 55
   DK-7184 Vandel

2. The Government Research Station
   Institute of Potatoes
   Tylstrup
   DK-9380 Vestbjerg

3. National Research Center for Plant Protection
   Lottenborgvej 2
   DK-2800 Lyngby

France

> Institut National de la Recherche Agronomique
> Station d'Amélioration de la Pomme de Terre et des Plantes à Bulbes
> 29207 Landerneau

Greece

> Ministry of Agriculture of Greece
> Variety Research Institute of Cultivated Plants
> Sindos – Thessaloniki

Ireland

1. The Agricultural Institute
   Oak Park Research Centre
   Carlow

2. Department of Agriculture
   Agriculture House
   Kildare Street
   Dublin 2

3. University College Dublin
   Department of Plant Pathology
   Faculty of Agriculture
   Belfield
   Dublin 4

Italy

> Istituto Sperimentale per le Colture Industriali
> Via di Corticella 133
> 40129 Bologna

Netherlands

> Rijksinstituut voor het Rassenonderzoek van Cultuurgewassen
> (RIVRO)
> Nieuwe Wageningseweg 1
> 6721 ND Bennekom

United Kingdom

1. Department of Agriculture & Fisheries for Scotland
    Agricultural Scientific Services
    East Craigs
    Edinburgh EH12 8NJ
    Scotland

2. Scottish Crop Research Institute
    Pentlandfield
    Roslin
    Midlothian EH25 9RF
    Scotland

3. National Institute of Agricultural Botany
    Huntingdon Road
    Cambridge CB3 0LE
    England

4. Northern Ireland Ministry of Agriculture
    Plant Breeding Division
    Loughall
    Co. Armagh BT6 18JB
    Northern Ireland

5. Northern Ireland Ministry of Agriculture
    Newforge Lane
    Belfast BT95 PX
    Northern Ireland

FR Germany

1. Institut für Pflanzenbau und Pflanzenzüchtung
    Bundesforschungsanstalt für Landwirtschaft (FAL)
    Bundesallee 50
    3300 Braunschweig

2. Varieties only on 'Beschreibende Sortenliste für Kartoffeln 1982'
    etc.
    Details of the breeder or agent who should be contacted to obtain
    seed stocks of a particular variety are given in the 'Sortenliste'

Descriptive references

For each variety details of one or two publications are listed, in which a botanical description, agronomic quality and disease characteristics are given for that variety. For a large number of varieties no descriptive reference is available from the reference sources at our disposal.

These publications are listed and given a reference number as follows:

1. 'Kartoffel – Nyt' No. 17 Sortsbeskrivelser

   Available from
   Kartoffelafgiftsfonden
   Tvaekaj 4, Trafikhavnen
   DK-6700 Esbjerg
   Denmark

2. Variety Descriptions

   Available from
   Danish Potato Breeding Foundation
   The Breeding Station
   Grindstedvej 55
   DK-7184
   Denmark

3. Statens Forsogsvirksomhed i Plantkultur, 386 Beretning, Beskrivelse of kartoffelsorter dyrket i Denmark, Tidsskrift for Planteavl 49: 1945

   Available from libraries and
   Bureau of Government Crop Husbandry Research Service
   Virumgaard
   Kongevejen 83
   DK-2800 Lyngby
   Denmark

4. 'Bulletin des Varietés Pommes de Terre' 1977 – 1983 etc.

   Published by
   Ministère de l'Agriculture et du Développement Rural

8 *Descriptive references*

Institut National de la Recherche Agronomique
Station d'Amélioration de la Pomme de Terre et des Plantes à Bulbes
29207 Landerneau
France

5. La Pomme de Terre (R. Diehl 1938), Charactères et Description des variétés, Monographies Publiées par les Stations et Laboratoires de Recherches Agronomiques, Ministère de l'Agriculture, France

6. Die Deutschen und Ausländischen Kartoffelsorten 1947/48, Schriftreihe für die Kartoffelwirtschaft (Dr. Siebencick 1947)

7. 'The Growing and Marketing of Potatoes for Seed'

Available from
Department of Agriculture
Agriculture House
Kildare Street
Dublin 2
Ireland

8. 'Potato Growing for Seed Purposes' by Davidson

Published in
Irish Department of Agriculture Journal Vol. XXXIV No. 2. 1936

9. Description of Potato Varieties bred by An Foras Taluntais

Available from
An Foras Taluntais
Oak Park Research Centre
Carlow
Ireland

10. Netherlands Catalogue of Potato Varieties 1985 etc.

11. Geniteurslijst voor Aardappelrassen 1981/82. Ing H. Zingstra

12. Beschrijvende Rassenlijst voor Landbouwgewassen 1985 etc.

   Publications 10, 11 and 12 available from
   RIVRO
   Postbus 32
   6700 AA Wageningen
   Netherlands

13. Classified List of Potato Varieties England and Wales 77/81 etc.

14. Unpublished Records of the National Institute of Agriculture

   Publications 13 and 14 available from
   National Institute Agricultural Botany (NIAB)
   Huntingdon Road
   Cambridge CB3 0LE
   England

15. Potato Varieties published by Potato Marketing Board and NIAB

   Available from
   NIAB
      or
   Potato Marketing Board
   50 Hans Crescent
   Knightsbridge
   London SWIX 0NB
   England

16. 'Seed Potatoes' – Department of Agriculture and Fisheries for Scotland 1956-61 etc.

   Available from
   Her Majesty's Stationery Office
   13A Castle St.
   Edinburgh 2
   Scotland

17. Potato Varieties – Redcliffe. N.Salaman M.D.

    Published by
    Cambridge University Press
    Fetter Lane
    London
    England

18. Beschreibende Sortenliste für Kartoffeln 1983 etc.

    Available from
    Bundessortenamt
    Osterfelddam 80
    3000 Hannover 61
    FR Germany

19. Kartoffelatlas (Deutsche Sorten)

    Published by
    'Die Kartoffelwirtschaft' GmbH
    Hamburg 1
    FR Germany

20. Polish Potatoes

    Published by
    Rolimpex
    Al. Jerozolimskie 44
    Warsaw
    Poland

21. Index 1982 of European Potato Varieties

    Available from
    Bibliothek der Biologischen Bundesanstalt für Land-und Forstwirt-
    schaft
    Königin-Luise-Strasse 19
    D-1000 Berlin 33 (Dahlem)
    FR Germany

22. Danish Potatoes

> Published by
> Kartoffeleksportudvalget
> Axelborg
> Copenhagen
> Denmark

23. Potato Handbook 1959. Potato Varieties Issue

> Published by
> The Potato Association of America
> New Brunswick
> New Jersey (Volume IV)
> USA

24. Seed Potatoes from Canada

> Published by
> The Department of Industry, Trade and Commerce
> Ottawa
> Canada

25. The Scottish Agricultural Colleges Potato Varieties Bulletin 1981

26. Sortenratgeber Kartoffeln 1963

27. Europäische und Nordamerikanische Sorten in Die Kartoffel, ein Handbuch. Band 11 (K.H.Moller 1962), R.Schick, M.Klinkowski, Berlin. Veb Deutscher Landwirtschaftsverlag

28. Weltkatalog der Kartoffelsorten, Verlag die Kartoffelwirtschaft, GmbH, Hamburg, (Dr. H.Siebeneck (1957))

29. Kartoffelsorten, Verlagsbuchhandlung, Paul Parey, Berlin, (Dr. Karl Snell (1929))

30. Centro Internacional de la Papa, Pathogen tested Potato Cultivars (May 1983)

31. Rassenbericht Aardappels, Variety bulletin for potatoes

   Available from
   RIVRO, Netherlands

32. Nederlands Rassenregister, Dutch Register of Varieties

   Technical information available from
   RIVRO, Netherlands

Members of Potato Working Group of the EC Plant Resistance and use of Gene Banks Programme Committee

Belgium
> Dr. G.Fouarge
> Station de Haute Belgique
> Rue de Serpont 48
> 6600 Libramont
> Chevigny

Denmark
> Dr. N.E.Foldo
> Landbrugets Kartoffelfond LKF
> Grindstedvej 55
> DK–7184 Vandel

France
> Dr. P.J.Perennec
> Station d'Amélioration de la Pomme de Terre
> 29208 Landerneau

Greece
> Dr. G.A.Andritos
> Ministry of Agriculture
> Plant Research Directorate
> Section of Horticultural Research
> G Kapnekoptiriou Str.
> Athens

Ireland
> Mr. H.W.Kehoe
> Plant Breeding Department
> An Foras Taluntais
> Oak Park Research Centre
> Carlow

Italy
> Dr. D.Cremaschi

Istituto Sperimentale per le Colture Industriali
Via di Corticella 133
40129 Bologna

Netherlands
Dr. Ir. H.Lamberts
Foundation for Agricultural Plant Breeding, SVP
P.O. Box 117
6700 AC Wageningen

Ir. M.J.Hijink
RIVRO
P.O. Box 32
6700 AA Wageningen

United Kingdom
Mr. David E.Richardson
National Institute of Agricultural Botany
Huntingdon Road,
Cambridge CB3 0LE England

Mr.G.MacKay
Scottish Crop Research Institute
Pentlandfield
Roslin, Midlothian EH25 9RF, Scotland

FR Germany
Frau Dr. G.Mix
Institut für Pflanzenbau und Pflanzenzüchtung
Bundesforschungsanstalt für Landwirtschaft (FAL)
Bundesallee 50
3300 Braunschweig

Commission Secretariat
J.Dehandtschutter
Commission of the European Communities DG VI F4
Rue de la Loi 200
1049 Brussels
Belgium
Mme. M.Wauters
Commission of the European Communities DG VI F4
Rue de la Loi 86
1049 Brussels
Belgium

List of named potato varieties
in alphabetical order

Variety	Belgium	Denmark	France	Greece	Ireland	Italy	Netherlands	United Kingdom	FR Germany	Descriptive reference	Synonym
Aba	–	–	–	–	–	D	–	–	–		
Abana	–	–	–	–	–	–	A	–	–	31,32	
Aberdeen Favorite	–	–	–	–	–	–	–	B1	–	14	
Abnaki	–	–	A	–	–	–	A	–	–	11,24	
Achat	–	–	A	–	–	–	A	–	A2	18,19,11,21	
Achievement	–	–	–	–	B2	–	–	B1	–	8	
Ackersegen	–	B1	A,C	–	B1,2	–	A	A1	D1	1,3,4,11,21	
Acrema	–	–	–	–	–	–	A	A2,B1	–	31,32	
Acresta	–	–	A	–	–	–	A	–	–	31,32	
Activa	–	–	–	–	–	–	A	–	–	32	
Adagio	–	–	–	–	–	–	A	–	–	32	
Adams Apple	–	–	–	–	–	–	–	B1	–		
Adelheid	–	–	–	–	–	–	A	–	–	11	
Adema	–	–	–	–	–	–	A	B1,C1	–	31,32	
Adirondack	–	–	–	–	B2	–	–	A1	–	8,16,17	

Variety									Merkur
Admirandus	—	—	A	—	—	—	A	—	11
Adonis	—	—	A	—	—	—	A	—	11,31
Adorina	—	—	—	—	—	—	A	—	31,32
Adretta	—	—	—	—	—	—	—	A3	—
Advira	—	B1	B	—	—	—	B	—	11
AEggeblomme	—	B1	—	—	—	—	—	—	—
Afke	—	—	—	—	—	—	A	—	32
Agitato	—	—	—	—	—	—	A	—	32
Agnes	—	—	D	—	—	—	—	A2	18,19,11,21
Agronomitsjeski	—	—	D	—	—	—	B	—	11
Aguti	—	—	A	—	—	—	A	A2	18,11,21,32
Ailsa	—	—	—	—	—	—	A1,2,3,5 C1	—	—
Ajax	—	—	A	—	—	C	A	—	10,11,12,4
Akebia	—	B1	B	—	—	—	—	D1	28
Aladdin	—	—	A	—	—	—	A	—	11,21
Alamo	—	—	—	—	—	—	A	—	11
Alaska Frostless	—	—	—	—	—	—	B	—	11
Alava	—	—	—	A1,2	—	—	—	—	21
Alannah	—	—	—	—	—	—	B1	—	8

Variety	Belgium	Denmark	France	Greece	Ireland	Italy	Netherlands	United Kingdom	FR Germany	Descriptive reference	Synonym
Albaro	–	–	–	–	–	–	A	–	–	31	
Alberta	–	–	D	–	–	–	–	B1	–	11,21	
Albion	–	–	–	–	–	–	B	B1	–	11	
Alcmaria	A	–	A	–	–	–	A	B1	D1	10,11,12,4, 31	
Aldo	–	–	–	–	–	–	A	–	–	31,32	
Alexa	–	–	–	–	–	–	–	–	D1	18,11	
Alhamra	–	–	–	–	–	–	–	B1,C1	–		
Alina	–	B1	–	–	–	–	–	–	–		
Alka	–	–	–	–	–	C	–	–	–		
Allard	–	–	–	–	–	C	A	–	–	32	
Allegro	–	–	–	–	–	–	–	–	A2	18,11,21	
Allerfruheste Gelbe	–	B1	D	.–	A1,2	–	A	A1	–	3,10,11,12	Palogan
Ally	–	–	–	–	–	–	–	A1	–	16,17,8	
Alma	–	B1	D	–	–	–	B	–	–	3,11,21,22	

Variety											
Almere	–	31,32	–	–	A	–	–	–	–	–	–
Alnwick Castle	Guardian	17	–	–	–	–	–	–	–	See Guardian	–
Alpha	–	10,11,12,1,7	–	A1,2,3,B4,C4	A	C	A1,2	–	A	B1,2,C3	–
Altena	–	31,32	–	A1,3,C1	A	–	–	–	–	–	–
Althena	–		–	–	–	–	C3	–	–	–	–
Alwin	–	32	–	–	A	–	–	–	–	–	–
Amalfy	–	11	D1	–	A	–	–	–	A	–	–
Amalia	–	18,11,21	A2	–	A	–	–	–	A	–	–
Amaranta	–	31,32	–	–	A	–	–	–	A	–	–
Amaryl	–	11,31,32	–	–	A	–	A1,2	–	D	B1	–
Amasa	–	31,32	–	–	A	–	–	–	–	–	–
Amazone	–	12,31	–	–	A	C	–	–	–	–	–
Ambassadeur	–	11,3,7,14,21	–	B1	A	–	A1,2,C3	–	A	–	–
Amber	O.P.Amber	9,11,13,21	–	A1,2,3	–	–	A1,2,C3	·–	A	–	B1
Ambra	–	11	C1	–	–	–	–	–	A	B1	–
Amedo	–	31,32	–	A2,B1	A	–	–	–	–	–	–
Amelio	–	11,19	–	–	A	–	A1,2	–	D	–	–
Amera	–	11,12,21,31	–	–	A	–	–	–	–	–	–
America	Irish Cobbler	8,14,16,17	–	A1	–	–	B2	–	–	–	–

Variety	Belgium	Denmark	France	Greece	Ireland	Italy	Netherlands	United Kingdom	FR Germany	Descriptive reference	Synonym
American Variety	–	–	–	–	B2	–	–	–	–		
Amethyst	–	B1	A	–	–	–	A	–	D1	11,18,21,31,32	
Amex	–	B1	B	–	–	–	A	–	–	11,19	
Amia	*See Amigo*									21	Amigo
Amigo	A	B1,2,C3	A	–	–	–	A	–	–	10,11,12,1,4,31	Amia
Aminca	–	–	A	–	–	C	A	A1,3,C1	–	10,11,12,4,31	
Amires	–	–	A	–	–	–	–	–	–		
Amred	–	–	–	–	–	–	A	–	–	31,32	
Amsel	–	B1	D	–	–	–	B	–	–	11,21	
Amva	–	B1	–	–	–	–	–	–	–	2,11	
Anco	–	–	–	–	–	–	A	–	–	11	
Andante	–	–	–	–	–	–	A	–	–	32	
Anett	–	–	D	–	–	–	–	–	D1	19,11	

Variety										
Angelika	–	–	–	–	–	–	–	D1		
Angus Leader	–	–	–	–	–	–	A1	–	16	
Aniel	–	A	–	–	–	A	–	–	4,11	
Anita	B1,D3	–	–	–	–	A	–	–	11	
Annabell	–	A	–	–	–	A	A1	D1	11,4,14,21	
Anosta	B1	A	–	B1	–	A	–	–	10,11,12,21, 31	
Ansje	–	–	–	–	–	–	–	D1	11	
Antar	–	–	–	–	–	–	B1,C1	–		
Antares	–	A	–	–	–	A	–	–	11	
Antigo	–	–	–	–	–	B	–	–	11	
Apache	–	–	–	–	–	–	A1,C1	–		
Apapit	–	–	–	–	–	–	–	D1	11,18	
Aphrodite	–	–	–	–	C	A	–	–	32	
Apis	–	–	–	–	–	–	–	D1	19	
Apolia	–	D	–	–	–	–	–	–		
Apollo	–	A	–	–	–	B	B1	D1	4,1,11,12,21, 10	Apollonia
Apollonia	*See* Apollo	–	–	–	–	–	–	–	21	Apollo
Apta	–	D	–	–	–	–	A1,3	C1	19,11	
Aquila	B1	D	–	A1,B2	–	A	B1	D1	19,11,14,21	

Variety	Belgium	Denmark	France	Greece	Ireland	Italy	Netherlands	United Kingdom	FR Germany	Descriptive reference	Synonym
Aranyalma	–	–	–	–	–	–	–	–	D1	11,21	
Arcona	–	–	–	–	–	–	A	–	–	31,32	
Arensa	–	–	B	–	–	–	–	–	D1	19,11	
Argo	–	–	–	–	–	–	B	–	–	11	
Argula	–	–	–	B	–	–	–	–	–		
Argyll Favourite	–	–	–	–	–	–	–	B1	–	14	
Ari	–	–	A	–	–	–	A	–	–	11	
Ariosa	–	–	–	–	–	–	A	–	–	32	
Ariane 74F-14-10	–	–	–	B	–	–	–	–	–		
Aristo	–	B1	–	–	–	–	A	–	–	11	
Arjan	–	–	A	–	–	–	A	–	–	11,21,31,32	
Arka	–	–	A	–	A1,2	C	A	A2,B1	–	10,11,12,4	
Arkula	–	–	A	–	–	C	A	A1,3,C1	–	10,11,12,13, 21	
Armen	–	–	A	B	–	–	–	–	–	4,11,21.	

Variety									D1	
Arnica	–	–	A	–	–	–	A	–	–	11
Arno	–	–	–	–	–	–	A	–	–	32
Aronia	–	–	–	–	–	–	–	A1	–	11
Arran Banner	–	B1	A,C	–	A1,2,C3	–	A	A1,2,3,5 C1,4	–	13,15,16,4,7
Arran Banner (Precose)	–	–	A	–	–	–	–	–	–	11
Arran Bard	–	–	D	–	–	–	–	B1	–	16
Arran Cairn	–	–	–	–	A1,B2	–	–	B1	–	14,16,8
Arran Chief	–	–	D	–	B2	–	–	A1,3,5	–	13,16,17,8
Arran Comet	–	–	–	–	A1,2	–	–	A1,2,3, C1	–	13,15,16,7
Arran Comrade	–	–	–	–	–	–	–	A1	–	14,16,17,8
Arran Consul	–	–	D	–	A1,2,C3	–	A	A1,2,3,5 B4,C1	–	13,15,16,7,8
Arran Crest	–	–	–	–	B2	–	B	A1	–	14,16,8,11
Arran Luxury	–	–	–	–	–	–	–	A1	–	14,16
Arran Peak	–	–	D	–	A1,2	–	–	A1,2,3	–	13,16,7,8,11
Arran Pilot	–	–	D	–	A1,2	–	A	A1,2,3,5 C1	–	13,15,16,7,8 11
Arran Rose	–	–	–	–	B2	–	–	A1	–	16,17,8

Variety	Belgium	Denmark	France	Greece	Ireland	Italy	Netherlands	United Kingdom	FR Germany	Descriptive reference	Synonym
Arran Scout	–	–	–	–	B2	–	–	B1	–	16,8	
Arran Signet	–	–	–	–	A1	–	–	A1	–	14,16,8	
Arran Victory	–	B1	D	–	A1,2,C3	–	B	A1,2,3,5 – B4	–	13,14,15,16 11	
Arran Viking	–	–	–	–	A1,B2	–	–	A1,3	–	13,16,11,21	
Arsy	–	–	–	–	–	C	A	B1	–	31,32	Mansholt 67–295
Aryo	–	–	D	–	–	–	–	–	–	11,14	
As	–	–	A	–	–	–	–	B1	–	11	Aspotet
Asaja	–	–	–	–	–	–	A	–	–	31,32	
Asche Samling	–	–	–	–	B2	–	A	–	–	23,14	
Ashworth	–	–	D	–	–	–	–	–	–	11,14,19	
Asoka	–	–	A	–	–	–	A	–	D1	1,3,11,21	
Asparges	–	B1,2,C3	–	–	–	–	–	–	–	18,11,21	Ratte
Assia	–	–	–	–	–	–	–	–	A2		

Variety										
Astarte	10,11,12,4, 31	–	–	A	–	–	–	A	–	–
Astilla	11,21	–	–	B	–	–	–	–	–	–
Astra	11,21	–	–	–	–	B1,2	–	–	–	–
Astrid	18,19,11,21	A2	–	–	–	–	–	B	–	–
Atacama		–	A2	–	–	–	–	–	–	–
Atara	32	–	–	A	–	–	–	–	–	–
Athene	19,3,11	–	–	–	C	–	–	–	B1	–
Atica	18,11,21,32	A2	B1	A	–	–	–	A	–	–
Atlas	11	D1	–	A	–	–	B	–	–	–
Atlantic		–	–	–	–	–	B	–	–	–
Atleet	11	–	–	A	–	A1,2	–	–	–	–
Atrela	12,31	–	A2,B1	A	–	–	–	–	–	–
Atzimba	30	–	A2	–	D	–	–	A,C	–	–
Augusta	19,11	D1	–	–	–	–	–	–	–	–
Aula	18,11,21,31, 32	–	B1	A	–	–	–	–	–	–
Aura	4,11,21	–	A1	A	–	–	–	A	–	–
Aurelia	11,19,21	–	–	B	–	–	–	–	–	–
Auriga	11	–	–	–	–	–	–	A	–	–

Variety	Belgium	Denmark	France	Greece	Ireland	Italy	Netherlands	United Kingdom	FR Germany	Descriptive reference	Synonym
Aurora	–	–	A	–	–	–	A	–	–	10,11,12,21, 31	
Ausonia	–	–	A	–	A2,D3	C	A	A1,3,C1	–	11,12,4,21, 31,10	
Avanti	–	B1	A	–	–	C	A	–	–	11,1,4,21	
Avenir	–	–	A	–	A2,B1	–	A	A1	D1	11,14,21	
Avondale	–	–	–	B	A1,2,C3	–	–	A1,3	–	9	
Axilia	–	–	A	–	–	–	A	–	–	11,21	
Azalia	–	–	–	–	–	C	–	–	–		
Babanki	–	–	A	–	–	–	–	–	–		
Baillie	–	–	–	–	A2	–	–	A1,2,3,5, C1	–	13,11,21	
Bake King	–	–	–	–	–	–	B	–	–	11	
Baku	–	–	A	–	–	–	–	–	–	19,11	
Balder	–	–	–	–	–	C	A	–	–	32	

Variety									
Ballinrobe Dutch Blue	–	–	–	B2	–	–	–	–	
Ballydoon	–	D	–	A1,2,C3	–	–	A1,3,5	–	13,16,11,8
Baltyk	–	–	–	B2	–	–	–	–	11
Bamberger Hornle	–	–	–	–	–	–	–	D1	
Baraka	–	A	–	–	C	A	A1,C1	D1	10,11,12,4
Barbara	–	–	–	–	–	–	B1	A2	18,11,21
Barima	–	A	–	B1	–	A	B1	–	11,14,21
Bartina	–	–	–	–	–	A	–	–	32
Bato	–	D	–	–	–	B	–	–	11
Bea	B1	A,C	–	B1	D	A	A2	–	10,11,12,4
Beauties	–	–	–	B2	–	–	–	–	8
Beauty of Bute	–	–	–	B2	–	–	B1	–	16,17,8
Beauty of Hebron	–	–	–	B2	–	–	–	–	23,8,17
Beko	–	–	–	–	–	B	–	–	11,21
Belaja Rosa	–	D	–	–	–	–	–	–	
Belchip	B1	–	–	–	–	–	–	–	
Belgium Lilly	–	–	–	B2	–	–	–	–	
Belita	–	–	–	–	–	A	–	–	31,32
Bella	–	–	–	–	–	B	–	–	11
Belladonna	–	A	–	–	–	–	–	A2	18,11,21

Variety	Belgium	Denmark	France	Greece	Ireland	Italy	Netherlands	United Kingdom	FR Germany	Descriptive reference	Synonym
Bellahouston	–	–	–	–	–	–	–	B1	–	14	
Belle Amelie	–	–	D	–	–	–	–	–	–		
Belle de Fontenay	–	–	A	–	–	–	B	–	–	4,11,21	
Belle de Locronan	–	–	A	–	–	–	–	–	D1	11	
Bellona	–	–	A	–	–	–	A	–	–	11,21,31	
Belorusskij Rannij	–	–	D	–	–	–	–	–	–	11	
Ben Cruachan	–	–	–	–	–	–	–	B1	–	17,8	
Benimaru	–	–	–	–	–	–	B	–	–	11	
Ben Lomond	–	–	–	–	B2	–	–	A1	–	16,8	
Benol	–	–	–	–	–	–	A	A2,B1	–	31,32	
Berber	–	–	–	–	–	C	A	B1,C1	–	31,32	
Berlichingen	–	–	–	–	–	–	B	–	D1	11	
Berlikummer Geeltje	–	–	–	–	–	–	A	–	–	11	
Bertita	–	–	–	–	B1	–	–	–	–		
Berolina	–	B1,2,C3	A	–	–	–	A	B1	A2	18,1,11,21	

Variety										
Beryl	—	—	—	—	—	D	—	—	—	—
Beta	—	—	—	—	—	—	—	C1	11	—
Bevelander	—	B1	A,C	—	—	—	A	—	11,3	—
B.F.15	—	—	A	—	—	—	A	A1	4,11,14,21	—
Bieka	—	—	—	—	—	—	—	B1	—	—
Biene	—	—	—	—	—	—	—	D1	—	—
Bildtstar	—	—	—	—	—	—	A	—	12,31	—
Binia	—	—	D	—	—	—	A	—	11	—
Binova	—	—	A	—	—	—	A	—	4,11,21	—
Bintje	A	B1,2,C3	A,C	—	A2,B1	C	A	A1,2,3,5 D1 / C1	10,11,12,1,3 13,15,16,20	Iturriet Temprana
Bintje Kaempetop	—	B1	—	—	—	—	—	—	—	—
Bintje × 3	—	—	—	—	—	—	B	—	—	—
Birga	—	—	—	—	—	—	—	D1	11	—
Bishop	—	—	—	—	B2	—	—	A1	16,17,8	—
Bison	—	—	A	—	—	—	—	D1	19,11	—
Black Bishop	—	—	—	—	—	—	—	B1	14	—
Black Castle	—	—	—	—	—	—	—	B1	14	—
Black Champion	—	—	—	—	B2	—	—	—	8	—
Black Kidney	—	—	—	—	—	—	—	A1	—	—

Variety	Belgium	Denmark	France	Greece	Ireland	Italy	Netherlands	United Kingdom	FR Germany	Descriptive reference	Synonym
Black King	–	–	–	–	–	–	–	A1	–	14	
Black Knight	–	–	–	–	–	–	–	B1	–	14	
Black Queen	–	–	–	–	–	–	–	A1	–	14	
Black Potato	–	–	–	–	B2	–	–	–	–		
Black Skerry	–	–	–	–	B2	–	–	–	–	8	
Bla Kartoffel DDSF	–	B,C1	–	–	–	–	–	–	–		
Blanik	–	–	D	–	B2	–	B	–	–	11,21	
Blanka	–	–	A	–	A2,B1	C	A	A1,2,3,5 C1	–	10,11,12,13	
Blauwe Eigenheimer	–	B1	–	–	–	–	A	–	–	11	
Blight Resistor	–	–	–	–	B2	–	–	–	–	8	
Bliss Triumph	–	–	D	–	–	–	–	–	–	11	
Bloomers	–	–	–	–	B2	–	–	–	–	8	
Blue Catriona	–	–	–	–	–	–	–	A1	–		
Blue Gloss	–	–	–	–	B2	–	–	A1	–	8,16	

Blue Grey	–	–	–	–	–	B1	–	8,16	–
Blue Kidney	–	–	–	B2	–	–	–	–	–
Blue Neb	–	–	–	B2	–	–	–	8	–
Blue Potato	–	–	–	–	–	A1	–	–	–
Bobbie Burns	–	–	–	–	–	B1	–	16,8	–
Bodenkraft	–	A	–	–	–	–	B2,C1	18,19,11,21	–
Boezigs Gelbbluhende	–	–	–	–	–	–	D1	–	–
Bohms Mi Medfruhe	*See* Mittelfruhe	–	–	–	–	–	–	–	Mittelfruhe
Bola	–	A	–	–	A	–	D1	19,11	–
Bona	B1	D	–	B2	–	–	D1	19,11	–
Bonnet De Noirmoutier	–	D	–	–	–	–	–	5	–
Bonnie Dundee	–	–	–	–	–	A1	–	–	–
Bonte Desiree	–	A	–	–	A	A1,3,C1	–	11,13,15,21	–
Bonus	–	D	–	–	–	–	–	–	–
Boreas	–	–	–	–	–	A1	–	25	–
Bornia	–	–	B	C	A	–	–	32,31	–
Borodjanski	–	D	–	–	–	–	–	11	Borodian-skij
Boston Kidney	*See* Dargill Early	–	–	–	–	–	–	8,17	Dargill Early

Variety	Belgium	Denmark	France	Greece	Ireland	Italy	Netherlands	United Kingdom	FR Germany	Descriptive reference	Synonym
Bothwell	–	–	–	–	B2	–	–	–	–		
Brandaris	–	–	A	–	–	–	A	–	–	11	
Brasovean	–	B1	A	–	–	–	A	–	–	11	
Bravo	–	–	–	–	–	–	A	–	–	11,8	
Brda	–	–	–	–	–	C	–	–	–		
Brenta	–	–	A	–	–	–	–	–	–	19	Vestar
Brennragis	–	B1	–	–	–	–	–	–	–		
Bright	–	–	–	–	–	–	A	–	–	32	
Briljant	–	–	A	–	–	–	A	–	–	11,21	
Brio	–	–	A	–	–	–	A	A1	D1	11,13,21,31	
British Queen	–	–	D	–	A1,2,C3	–	–	A1,3,5, B4,C1	–	13,16,17,7,8	
Britta	–	–	–	–	–	–	–	B1	A2	18,11,21	
Broca	–	–	A	–	–	–	A	–	D1	19,11	
Bronderslev Kartoffel	–	B,C1	–	–	–	C	–	–	–	5	
Bronka	–	–	–	–	–	C	–	–	–		

Variety										Synonym
Brunella	–	–	A	–	–	–	–	C1	11	
Bryza	–	–	–	–	C	–	–	–		
Buchan Beauty	–	–	–	–	B2	–	–	B1	8,16	
Buffs	–	–	–	–	B2	–	–	–	8	
Burbank	–	B,C1	D	–	–	–	–	–	23.3	
Burmania	–	–	A	–	A1,B2	A	–	B1,4	11,14	
Bustan	–	–	–	–	–	–	–	B1,C1		
Bute Blue	–	–	–	–	–	–	–	A1		
Cadora	–	–	A	–	–	–	–	–		
Calori	–	B1	D	–	–	–	–	–	11	
Calrose	–	B1	–	–	–	–	–	–	23,11	
Calypso	–	–	–	–	–	–	–	A1,C1		
Canoga	–	–	–	–	–	–	–	B1	23,14	
Capella	–	B1	D	–	B1	A	–	–	11,21	Lenino
Capiro	–	–	A	–	–	–	–	–		
Cara	–	–	A	B	A1,2,C3	–	–	A1,2,3,5 C1,4,C4	9,11,13,15	O.P.Beauty
Cardinal (NL)	–	–	A	–	C	A	–	A1,2,3 C1	10,11,12,4,8 31	

Variety	Belgium	Denmark	France	Greece	Ireland	Italy	Netherlands	United Kingdom	FR Germany	Descriptive reference	Synonym
Cardinal (GB)	–	–	–	–	–	–	–	A1	–	–	
Capri	–	–	A	–	–	–	–	–	–	–	
Caribe	–	–	–	–	–	C	–	–	–	–	
Carina	–	–	A	–	B1	–	A	A1,2,5	A2	11,12,18	
Carla	–	–	A	–	–	–	–	–	D1	19,11	
Carlingford	–	–	–	–	A2	–	–	A1,3,5, C1	–	13	
Carmen	–	B1	–	–	–	–	–	–	D1	19,11	
Carmen Tcheque	–	–	D	–	–	–	–	–	–	6	
Carnea	–	B1	D	–	–	–	–	–	D1	11	
Carola	–	–	–	–	–	–	–	–	A2	18,11,21	
Carpatin	–	–	D	–	–	C	–	–	–	11	
Cascade	–	–	–	–	–	–	A	–	–	11	
Caspar	–	B1	–	B	–	–	A	–	–	32	
Catarina	–	–	A	–	A1,2	–	A	–	–	4,11,21	

Variety										
Cati	–	–	–	–	–	C	–	–	–	–
Catriona	–	–	–	–	A1,2,C3	–	–	A1,2,3, B4,C1	–	13,16,17,7,8
Cayuga	–	B1	–	–	B1	–	A	–	–	23,11,14
Cellini	–	–	D	–	–	–	–	–	–	5
Celt	–	–	–	–	–	–	–	B1	–	14
Centifolia	–	–	–	–	–	–	–	B1	D1	11
Ceres	–	–	A	–	–	–	–	–	D1	17,11,31
Certa	–	–	–	–	–	C	–	–	–	–
Cetyrehsotka	–	–	D	–	–	–	–	–	–	
Champion	–	–	D	–	A1,2	–	–	A1,B4	–	16,17,7,8,11
Champion Heir	–	–	–	–	B2	–	–	–	–	8
Chancellor	–	–	A	–	–	–	–	A1,3	–	13,16,11,21
Charlotte	–	–	A	–	–	C	A	–	–	4,11,21,32
Chenango	–	B1	–	–	–	–	–	–	–	23
Cherokee	–	–	D	–	B1	–	B	–	–	23,11,24
Chippewa	–	–	–	–	–	–	A	–	–	23,11
Christa	A	–	A	–	–	C	A	A1,3,C1	A2	18,11,21,31, 32
Churchill	–	–	–	–	–	–	–	A1	–	
Cilena	–	–	–	B	–	C	–	–	A2	18,11,21

Variety	Belgium	Denmark	France	Greece	Ireland	Italy	Netherlands	United Kingdom	FR Germany	Descriptive reference	Synonym
Civa	–	–	D	–	B1	C	A	A1,C1	–	10,11,12,13	–
Clada	–	–	A	B	A1,2,C3	–	–	A1,3,C1	–	9,11,13,21	O.P.Bounty
Clan Donnachie	–	–	–	–	–	–	–	A1	–		
Claudia	–	B1	A,C	–	A1,2	–	A	A1	–	4,11,21,32	
Claustar	–	–	A,C	B	A1,2,C3	C	A	–	–	4,11,21	
Claymore	–	–	–	–	–	–	–	B1	–	16,8	
Cleopatra	–	–	A	–	A2,C3	C	A	A1,3,C1	–	10,11,12,13	
Climax	–	B1	B,D	–	A1,B2	C	A	A1,B4	D1	10,11,12,13	
Clivia	–	B1	A	–	–	–	B	–	B2,D1	18,19,11,21	
Clovullin	–	–	–	–	–	–	–	B1	–	14	
Cobra	–	–	D	–	–	–	A	–	D1	19,11	
Colina	–	B1	–	–	–	–	A	–	–	11,21	
Colmo	–	–	A	–	C3	C	A	A1,3,C1	–	10,11,12,4	
Colossal	–	–	–	–	B2	–	–	–	–	17,8	St. Malo Kidney

Variety										
Columba	–	–	D	–	–	–	–	–	–	28
Comle	–	B1	–	–	–	–	A	A1,C1	–	11,13
Commandeur	–	–	A	–	–	–	A	A1	–	11,4,14,21
Compagnon	–	–	A	–	–	–	A	–	–	11,4,21
Comtessa	–	–	–	–	–	–	–	–	D1	11,14
Conchita	–	–	–	–	–	–	–	–	D1	30
Concorde	–	–	–	–	–	–	–	A5,B1,C1	–	
Concurrent	–	–	–	B	–	C	A	B1,C1	–	32
Condea	–	–	A	–	–	–	B	–	A1,2	18,19,11,21
Condor	–	–	D	–	–	–	–	–	D1	19
Conference	–	–	A	–	–	–	–	A1,3	–	13,16,11,21
Congo	–	–	–	–	–	–	–	B1	–	14
Conny	–	–	–	–	–	–	B	–	D1	11,21
Conquest	–	–	–	–	–	–	–	B1	–	16
Constante	–	–	A	–	–	C	A	–	–	11,21,32
Contessa	–	–	–	–	–	–	–	A1,3	–	11,13,21
Cordia	–	B1	–	–	–	–	–	–	B1	11
Corine	–	–	A	–	A1,2	C	A	A1,2,5 C1	B2,D1	10,11,12,13

Variety	Belgium	Denmark	France	Greece	Ireland	Italy	Netherlands	United Kingdom	FR Germany	Descriptive reference	Synonym
Corona	–	–	–	–	–	–	–	–	D1	19,8,11	–
Coronation	–	–	–	–	–	–	–	A1	–	14,17,8	–
Corrib	–	–	–	–	A1,2	–	–	–	–	9,11,21	O.P. Avenger
Corrie	–	–	–	–	–	–	–	A2	–	14	
Corsair	–	–	–	–	–	–	–	A2,3	–		
Cosima	–	–	–	–	–	–	–	–	B2,C1	18,19,11,21	
Costa	–	–	–	–	–	C	–	–	–		
Costella	–	–	–	–	–	–	–	A1,3,C1	–		
Craigneil	–	–	–	–	–	–	–	A1	–		
Craigs Alliance	–	–	–	–	A1,2	–	–	A1,2,3,5 C1	–	13,15,16,7	
Craigs Bounty	–	B,C1	–	–	B1,2	–	–	B1	–	14,16,11	
Craigs Defiance	–	–	A	–	A1,2	–	A	A1,2	–	14,16,11	
Craigs Royal	–	–	–	–	A1,2	–	A	A1,2,3	–	13,16,7,11	

Craigs Snowhite	–	–	D	–	B1	–	–	A1,2,3,4	–	14,16,11
Creata	–	–	–	–	–	–	A	–	–	32
Cresus	–	–	D	–	–	–	–	–	–	11
Crimson Beauty	–	–	–	–	–	–	–	A1	–	17,8
Cromwell	–	–	–	–	–	–	–	A1,3,C1	–	
Croft	–	–	A	–	A2,B1	–	–	A1,2,3,5 C1	–	13,15,11,21
Crusader	–	–	D	–	B2	–	–	B1	–	16,17,8,11
Cruza	–	–	–	–	–	–	–	A2	–	
Culpa	–	–	A	–	–	–	A	B1	A2	18,11,21,31
Cumnock	–	–	–	–	–	–	–	B1	–	14
Curran Dutch Potato	–	–	–	–	B2	–	–	–	–	
Cvetnik	–	–	–	–	–	–	A	–	–	11,21
Cynia	–	–	–	–	–	C	–	–	–	
Czarine	–	–	D	–	–	–	–	–	–	5
Dagmar	–	–	D	–	–	–	–	–	–	6
Dakota Red	–	–	D	–	–	–	–	–	–	11,14
Dalby Special	–	–	–	–	–	–	–	B1	–	
Dalco	–	–	D	–	B1	–	A	–	–	11

Variety	Belgium	Denmark	France	Greece	Ireland	Italy	Netherlands	United Kingdom	FR Germany	Descriptive reference	Synonym
Dalia	–	–	–	–	–	C	–	–	–		
Danae	A	–	A	–	–	–	A	–	–	4,11,21	
Dani	–	–	A,C	–	–	–	A	–	–	4,11,21	
Danva	–	B1,2,C3	A	–	A2,C3	–	A	A1,3,C1	–	1,2,11,13,21, 32	
Daresa	–	–	A	–	A1,2	–	A	–	–	10,11,12,4	
Daresa Rouge	–	–	A	–	–	–	–	–	–		
Daresa Wilding	–	–	–	–	B2	–	–	–	–		
Dargill Early	–	–	–	–	–	–	–	A1	–	14,16,17	Boston Kidney
Daroli	–	–	A	–	–	–	A	–	–	11	
Darwina	–	–	–	–	–	C	A	A1,2,3, C1,4	–	11,12,21,31	
Datcha	–	–	A	–	–	–	A	–	–	11	Datscha
Datscha	*See* Datcha										Datcha
Datura	–	B1	A	–	B1	–	–	A1	A2	18,19,11,12	

										Datura
Datura Kameke	–	–	D	–	–	–	–	–	–	27
Dazoc	–	–	–	–	–	–	–	–	C1	23,11
Dean Wilson	–	–	–	–	–	–	–	B1	–	–
Debora	–	–	D	–	–	–	A	–	–	11
Dekama	–	–	–	–	A1,2	–	–	–	–	11,7,21
Delcora	–	–	–	–	–	C	A	–	A5,B1,C1	32
Delica	–	–	–	–	–	–	–	–	A2	18,11,21
Delos	–	B1	D	–	–	–	–	–	C1	19,11
Denali	–	B1	–	–	–	–	–	–	–	–
Deodara	–	B1	D	–	–	–	B	B1	–	11
Depesche	–	–	–	–	–	–	–	–	D1	11,19
Desiree	A	B1,2,C3	A,C	–	A1,2,C3	C	A	A1,2,3,5	B2,D1 C1,4	10,11,12,1,4
Detskoselskij	–	–	D	–	–	–	–	–	–	–
Deva	–	–	D	–	–	–	–	–	–	11
Dextra	–	–	–	–	–	–	–	–	D1	11,31
Diamant	–	–	A	B	–	C	A	–	A1,3,C1	10,11,12,21,31
Diana	–	–	A	–	–	C	A	–	A1,3,C1	11,13,21,32
Dianella	A	B1,2,C3	A	B1	–	–	B	–	–	1,11,14,21

Variety	Belgium	Denmark	France	Greece	Ireland	Italy	Netherlands	United Kingdom	FR Germany	Descriptive reference	Synonym
Digna	–	–	A	–	–	–	–	–	D1	19,11	
Dir Johannssen	–	–	D	–	–	–	–	–	–	5	
Dirus	–	–	–	–	–	–	–	–	D1	11	
Diva	–	–	A	–	–	–	–	–	–	14	
Di Vernon	–	–	D	–	A1,2	–	B	A1,3,C1	–	13,16,17,8,11	
Dobbies Asset	–	–	–	–	–	–	–	B1	–		
Dobrin	–	–	–	–	–	–	A	–	–	11,21	
Domeka	–	–	–	–	–	–	A	–	–	31,32	
Domina	–	–	–	–	–	–	–	–	A2	18,11,21	
Dominion	–	–	–	–	–	–	⌐	B1	–		
Dona	–	–	–	B	–	–	–	–	–		
Donard	–	–	–	–	–	–	–	A1	–	8	
Donard Nursery	–	–	–	–	B2	–	–	–	–		
Donata	–	–	A	–	–	–	A	–	–	11,32	
Donor	–	–	A	–	–	–	–	–	–	11	

Variety											
Doon Bounty	–	–	–	–	–	–	–	–	B1	–	11
Doon Castle	–	–	–	–	–	–	–	–	B1	–	16
Doon Cavalier	–	–	–	–	B1	–	–	–	–	–	16,8
Doon Early	–	–	–	–	B1,2	–	–	–	A1	–	16,11
Doon Eire	–	–	–	–	B1	–	–	–	B1	–	8
Doon Pearl	–	–	D	–	–	–	–	–	A1	–	
Doon Star	–	–	D	–	B1,2	–	–	–	A1,2,3, B4,C1	–	13,16,8,11
Doon Well	–	–	–	–	–	–	–	–	B1	–	16
Dora	–	–	B	–	–	–	–	–	–	–	11
Dorado	–	–	–	–	–	–	–	A	–	–	11
Doranda	–	–	–	–	–	–	–	–	·	D	
Dore	B	B1,C3	D	–	–	–	–	A	–	–	10,11,12,21
Dorita	–	B1	–	–	B1	–	–	A	–	–	11
Draga	–	B1	A	–	–	–	C	A	A1,3,C1	–	10,11,12,4
Draga Tcheque	–	–	A	–	–	–	–	–	–	–	
Drayton	–	–	A	–	A1,2	–	–	A	A1,2,3,5 C1	–	13,15,11,21
Dr McIntosh	–	B1	–	–	–	–	–	A	A1,2,3, B4,C1	–	13,16,7,11
Drossel	–	–	D	–	–	–	–	–	–	–	11,27

Variety	Belgium	Denmark	France	Greece	Ireland	Italy	Netherlands	United Kingdom	FR Germany	Descriptive reference	Synonym
Drummond Castle	–	–	–	–	–	–	–	B1	–	13	
Druzba	–	–	D	–	–	–	–	–	–		
Duchesse	–	–	D	–	–	–	–	–	–	5	
Duet	–	–	–	–	–	C	–	–	–		
Duivelander	–	–	D	–	–	–	A	–	–	11	
Duke of Kent	–	–	–	–	–	–	–	B1	–	16,8	
Duke of York	–	–	–	–	A2,C3	–	–	A1,2,3, C1	–	13,15,16,17	Eersteling
Dunbar Archer	–	–	–	–	–	–	–	B1	–	16,8	
Dunbar Cavalier	–	–	–	–	B2	–	–	B1	–	8,11	
Dunbar Rover	–	–	D	–	A1,2,C3	–	A	A1,3,5, B4	–	13,16,7,8	
Dunbar Standard	–	–	–	–	A1,2,C3	–	–	A1,2,3,5, B4,C1	–	13,15,16,8	
Dunbar Yeoman	–	–	D	–	B1,2	–	–	A1,2,3,- C1	–	16.8.11	
Dundrum	–	–	–	–	–	–	–	A1,3,5, C1	–		

Variety											
Dunluce	–	–	–	–	–	A2	–	A1,3,5	–	13,15,11,21	
Dunnottar Castle	–	–	–	–	–	–	–	A1	–	8	
Earlaine	–	B1	D	–	–	–	–	–	D1	23,11	
Earliest of All	–	–	D	–	–	–	–	–	–	14	
Earl of Essex	–	–	–	–	–	B2	–	–	–	8	
Early Gem	–	–	–	–	–	–	B	–	–	23,11	
Early Market	–	–	–	–	–	B2	–	A1	–	16,17,8	
Early Ohio	–	–	D	–	–	B2	–	–	–	23,8,11,24	
Early Pink Champion	–	–	–	–	–	–	–	B1	–		
Early Regent	–	–	–	–	–	B2	–	–	–	8	
Early Rose	–	–	A,C	–	–	–	A	A1	–	23,4,8,11,16, 21,24	Tidlig Rosen
Early Templar	–	–	–	–	–	B2	–	–	–	8	
East Nevk	*See* Southesk										Southesk
Eba	A	B1	A	–	C	A1,2	A	A3,B1	–	10,11,12,4	
Echo	–	–	A	–	–	–	A	–	D1	19,11	
Eclata	–	–	B	–	–	–	–	–	–	11,31	
Eclipse	–	–	–	–	–	B1,2	–	A1,3	–	13,16,8,11,21	Sir John Llevelyn

Variety	Belgium	Denmark	France	Greece	Ireland	Italy	Netherlands	United Kingdom	FR Germany	Descriptive reference	Synonym
Edda	–	–	A	–	–	–	–	–	A2,D1	18,19,11,21	
Edelgart	–	–	D	–	–	–	–	–	–	6	
Edeltraut	–	–	D	–	–	–	–	–	–	5	
Edgecote Purple	–	–	–	–	B2	–	–	A1	–	8,16,17	
Edinburgh Castle	–	–	–	–	–	–	–	A1	–	17,8	
Edith	–	–	A	–	–	–	–	–	A2	18,11,21	
Edzell Blue	–	–	–	–	B2	–	–	A1,2,3, C1	–	13,16,17,8	
Edzina	–	–	A	–	–	C	A	A1	–	10,11,12,4	
Eersteling	–	–	–	–	–	–	A	–	–	13,10,12,21 22	Duke of York
Eerstelingen	–	–	A	–	–	–	–	–	–	14,4,5,11	Ersling Eersteling
Ehud	–	–	B	–	A1,2	–	A	–	–	11,12,21	
Eigenheimer	–	B1	A	–	A2,B1	–	A	A2,B1	–	11,12,8,21	
Eightyfold	–	–	–	–	B2	–	–	–	–	16,17,8	

Evergood

									Evergood
Eke	—	—	—	—	—	—	A	—	31,32
Elan	—	A	—	—	—	—	—	—	11
Eldorad	*See* Evergood								
Electre	B1	D	—	—	—	—	A	—	11
Eleisa	—	—	—	—	—	—	—	A2,B1	18,11,21
Elektra	—	—	—	—	C	—	A	—	32
Element	B1	A	—	A1	—	—	A	—	10,11,12,4,31
Elena	—	—	—	—	—	—	A	—	32
Elida	—	—	—	—	C	—	—	—	
Elipsa	—	—	—	—	C	—	—	—	
Elkana	—	A	—	—	—	—	A	—	11,12,21,31, 10
Elke	—	A	—	—	—	—	—	D1	19,11
Ella	—	—	—	—	—	—	—	D1	11
Elles	—	—	—	—	—	—	A	—	31,32
Elsa	—	D	—	—	—	—	—	—	11
Elvingston Gowan	—	—	—	—	—	A1	—	—	11,13,14
Elvira	A	B	—	—	D	—	A	D1	10,11,12,21 31
Emergo	—	A	—	—	—	—	A	—	11,12,21
Empire	—	—	—	—	D	—	—	—	

Variety	Belgium	Denmark	France	Greece	Ireland	Italy	Netherlands	United Kingdom	FR Germany	Descriptive reference	Synonym
Endra	–	–	A	–	–	–	–	–	–	19,11	
Endzina	–	–	–	B	–	–	–	–	–		
Entente Cordiale	–	–	–	–	–	–	–	B1	–	17	
Epicure	–	–	D	–	A1,2,C3	–	B	A1,2,3, C1	–	13,15,16,17	
Epoka	–	–	D	–	–	–	B	–	–	11,21	
Erato	A	–	D	–	–	–	–	–	–	11	
Erbium	–	–	A	–	–	–	–	–	D1	19,11	
Erdgold	–	–	D	–	–	–	B	–	–	19,11	
Erdkraft	–	B1	–	–	–	–	–	–	–	11	
Erdmanna	–	–	–	–	–	–	–	–	D1	19,11	
Erendira	–	–	–	–	B1	–	–	–	–		
Erna	A	–	A	–	–	–	–	–	D1	18,11,21	
Erntekrone	–	–	–	–	–	–	–	–	D1	11	
Erntesegen	–	–	–	–	–	–	–	–	D1	11	

Variety											
Erntestolz	–	B1	A	–	–	–	A	B1	A2	18,11,21,32	
Eroica	–	–	–	–	–	–	–	B1	–	–	
Erstling	–	B1,2,C3	–	–	–	–	A	–	D1	1,3,19,21	Eersteling
Erstling (D)		*See* Eerstelingen									
Escort	–	–	A	B	–	C	A	C1	–	11,4,21,31,32	
Esperante	–	–	–	–	–	–	A	B1	–	32	
Essex	–	–	–	–	–	–	–	B1	–	23,11	
Esta	–	–	A	–	–	–	–	–	A2,D1	18,11,21	
Estima	–	–	A	–	A2,B1	C	A	A1,2,3,5,– C1	–	10,11,12,4	
Etoile Du Leon	–	–	A	–	A1,2	–	B	A1,3	–	4,11,13,21	
Etoile Du Nord	–	–	–	–	–	–	–	B1	–	17	
Eureka	A	–	A,C	B	–	–	A	B1	–	4,11,21	
Eva	–	–	D	–	–	–	–	–	D1	18,19,11	
Evergood	–	–	D	–	B2	–	A	A1	–	16,17,8,11	Eldorad
Exita	–	–	D	–	–	–	–	–	–	11	
Exodus	–	B1	B	–	–	–	A	B1	–	10,11,12,21	
Exponent	–	–	–	–	–	–	A	–	–	31,32	
Expova	–	B1	–	–	–	–	–	–	–		
Extase	–	–	B	–	A1,B2	–	A	–	–	11	

Variety	Belgium	Denmark	France	Greece	Ireland	Italy	Netherlands	United Kingdom	FR Germany	Descriptive reference	Synonym
Fala	–	–	–	–	–	C	–	–	–		
Falke	–	–	D	–	–	–	–	A1	D1	19,11	
Fambo	–	–	–	B	–	D	A	A3,B1,C1	–	32,31	
Famosa	–	–	A	B	A2	C	A	A1,3,5,C1	–	11,12,4,13,10	
Fanal	B	B1	A	–	–	–	–	–	D1	18,19,11	
Fanette	–	–	–	B	–	D	–	–	–		
Fanfare	–	–	–	–	–	–	A	–	–	31,32	
Farfadette	–	–	A	–	–	–	A	–	–	4,11,21	
Farmers	–	–	–	–	B2	–	–	–	–	8	
Fatima	–	–	A	–	–	–	A	B1	D1	18,19,11,21	
Fauna	–	–	–	–	–	C	–	–	–		
Fausta	–	–	A	–	–	–	A	–	A2	18,11,21	
Favoriet (1)	–	–	C	–	–	–	–	–	–	5	
Favorita (1)	–	–	A	–	–	C	A	–	–	10,11,12,4	

Variety									
Fecula	–	A	–	–	–	–	–	–	11
Fecuva	–	B1,2,C3	–	–	–	–	–	–	2
Feja	–	B1	–	–	–	A	–	–	4,1,11,21
Feldeslohn	–	D	–	–	–	–	–	–	19,11
Feuergold	–	D	–	–	–	–	–	–	5
Fianna	–	–	–	–	–	A	–	–	31,32.
Field Marshall	–	–	–	B2	–	–	–	–	8,16,17
Fiftyfold	–	–	–	–	–	–	A1	–	16,8
Filli	–	–	–	–	–	–	–	D1	11
Fina	A	A	B	–	–	–	–	A2	18,19,11,21
Fink	–	D	–	–	–	–	–	–	11
Finn Early	–	–	–	B1	–	–	–	–	–
Fiona	–	–	–	B1	–	–	A1,2,3,5 C4	–	14,11
Firapple Pink	–	–	–	B2	–	–	–	–	13
Fitoftorovstojcivyi	–	D	–	–	–	–	–	–	–
Flamenco	–	–	–	–	C	A	–	–	32
Flamingskost	–	D	–	–	–	–	–	–	11
Flaminia	–	A	–	–	–	–	B1	–	11
Flava	B1	D	–	B2	–	B	–	D1	19,3,11

Variety	Belgium	Denmark	France	Greece	Ireland	Italy	Netherlands	United Kingdom	FR Germany	Descriptive reference	Synonym
Flisak	–	–	–	–	–	C	–	–	–	–	
Flora	–	–	A	–	–	–	A	–	–	11,21	
Florita	–	B1	–	–	–	–	–	–	–		
Flouder	–	–	–	–	B2	–	–	–	–		
Flourball	–	–	D	–	A1,2	–	–	A1,B2	–	8,11,16,17	
Fluke	–	–	D	–	–	–	–	–	–	14,5	
Foca	–	–	–	–	–	C	–	–	–		
Fontana	–	–	–	–	B2	–	B	–	–	11	
Forelle	–	–	A	–	–	–	A	–	B2,D1	18,19,11,21,32	
Foremost	–	–	–	–	–	–	–	A1,3,C1	–		
Format	–	–	A	–	A1,2	–	A	–	–	11,21	
Fortuna	–	B1	–	–	B1,2	–	A	–	B2,D1	18,19,11,21, 32	
Fortyfold	–	–	–	–	–	–	–	A1	–	8,16	
Foula Red	–	–	–	–	–	–	–	A1	–		Geante

Variety										
Fox	–	–	–	–	–	–	–	A2	8,11,21	
Foxton	–	–	–	A2,B1	–	–	A1,2,3,5,–C1	A2	13,11,21	
Fram	–	–	–	–	–	B	–	–	11	
Fransen	–	–	–	–	–	B	–	–	11	Jaune D'or
Franzi	–	A	–	–	–	–	–	A1,2	18,11,21	
Freia	–	–	–	–	–	–	–	D1	18,11	
Fresco	–	–	–	–	–	A	–	A1	31,32,12	
Frezja	–	–	–	–	C	–	–	–		
Frigga	–	A	–	–	–	A	–	D1	18,19,11,21	
Frila	–	A	–	–	–	A	–	A2	18,19,1,4,11,31	
Frisia	–	–	B	–	D	–	–	–		
Friso	–	–	–	–	–	A	B1	–	11	
Froma	–	A	–	B1	–	–	–	–	11	
Fronika	–	–	–	–	–	–	A1,3,C1	–		
Fruhbote	B1	D	–	–	–	–	–	D1	19,3,11	
Fruheste Delikatess	–	D	–	A1,B2	–	–	–	–	11	
Fruhmolle	B1	–	–	–	–	–	B1	–	11	
Fruhmolle Tardive	–	D	–	–	–	–	–	B1		

Variety	Belgium	Denmark	France	Greece	Ireland	Italy	Netherlands	United Kingdom	FR Germany	Descriptive reference	Synonym
Fruhnudel	–	–	D	–	–	–	–	–	D1	11	
Fruhperle	–	B1	A	–	B1	–	–	–	–	11	
Fruhgold	–	–	C	–	–	–	–	–	–	5	
Fuga	–	–	–	–	–	–	–	–	D1	11	
Fulunskij	–	–	D	–	–	–	–	–	–		
Fundi	–	–	–	B	–	–	–	–	–		
Furore	–	–	A	–	B1	–	A	B1	–	11,5	
Furore Framboise	–	–	D	–	–	–	–	–	–		
Gabi	–	–	–	–	–	–	A	–	A1	19,11	
Gabriela	–	B1	–	–	–	–	–	–	–	30	
Gallo	–	–	–	–	–	–	–	–	D1	19,11	
Gamma	–	–	A	–	–	–	–	–	D1	11	
Garant	–	–	–	–	–	D	–	–	–		
Garden Filler	–	–	–	–	B2,D3	–	–	B1	–	14	

Variety									
Gari	A	–	A	–	–	–	–	–	11
Garibaldi	–	–	–	–	A	–	–	–	32
Gaspar	–	–	–	–	–	D	–	–	–
Gatcinskii	–	–	–	–	B	–	–	–	11
Gaumaise	B	D	–	–	–	–	–	–	11,21
Gawkies	–	–	–	B2	–	–	–	–	8,17
Geante	*See* Fluke	–	–	–	–	–	–	–	Fluke
Geante Bleue	–	D	–	–	–	–	–	–	5
Geelblom	–	D	–	–	–	–	–	–	11
Gelbling	–	B	–	–	–	–	–	–	–
Gelda	A	–	–	–	A	–	–	A2	18,11,21,31,32
Gelderse Rode	–	–	–	–	B	–	–	–	11
Gemma	–	D	–	–	–	–	–	–	6
General	–	–	–	–	–	–	–	B1	14,17 Up to Date
Genevievre	A	D	–	–	–	–	–	–	–
Gibridryj	–	D	–	–	–	–	–	–	–
Gideon	–	D	–	–	–	–	–	–	11
Giewont	–	–	–	B2	A	–	–	–	20,11,21
Gigant	–	B	–	–	A	C	B	B1	31,32

Variety	Belgium	Denmark	France	Greece	Ireland	Italy	Netherlands	United Kingdom	FR Germany	Descriptive reference	Synonym
Gineke	–	–	D	–	B1,2	–	A	B1,4	–	11,21	
Gisele	A	–	D	–	–	–	–	–	–		
Gladblaadje	–	–	–	–	–	–	B	–	–		
Gladstone	–	–	A	–	A2,B1	–	–	A1,3	–	13,16,8,11	
Glasgow Favorite	–	–	–	–	–	–	–	B1	–	14	
Glenalmond	–	–	–	–	–	–	–	B1	–	14	John Bull
Glen Clova	–	–	–	–	–	–	–	B1	–	14	
Glenesk	–	–	–	–	–	–	–	A1	–	14	
Gleniffer	–	–	–	–	–	–	–	A1	–	14	
Glenshee	–	–	–	–	–	–	–	B1	–	14	
Gloria	–	–	A	–	–	C	A	–	A2	18,10,11,12,31	
Gloria (Aucuba)	–	–	–	–	–	–	B	–	–		
Gloria (NAK)	–	–	D	–	–	–	–	–	–	28	
Gloria Modrow	–	–	A	–	–	–	–	–	–	28	

Variety									
Golden	*See* Kathadin *and* Kathadin Hollandaise								
Golden Wonder	B1	D	–	A1,2,C3	–	A1,2,3,C1	–	13,15,16,17	Kathadin
Goldniere	–	–	–	–	–	–	D1		
Goldwahrung	–	D	–	–	–	–	–	6	
Goliath	–	A	–	B1	A	–	–	11	
Gracia	B1	A	–	C	A	A1,3,5	–	10,11,12,13	
Gracilia	B1	B	–	–	–	–	–		
Grandia	–	–	–	–	–	A1	–	14	
Grandifolia	–	A	–	–	–	B1	A2	18,11,21	
Granola	–	A	–	–	–	B1,C1	A2	18,11,21	
Grata	–	A	–	–	B	–	B2,D1	18,19,11,21	
Grazia	*See* Gusto								Gusto
Great Scot	–	D	–	A1,2	A	A1,3	–	13,16,17,8	
Green–Budded Kerr's Pink Rogue	*See* Kerr's Pink Green Bud								Kerr's Pink Green Bud
Green Champion	–	–	–	B2	–	–	–	8	
Green Mountain	–	D	–	B2	B	–	–	23,11,14,17	
Gregor Cups	–	–	–	B2	–	A1	–	8,16	
Greta	–	–	–	–	–	–	D1	14	
Guadelupe	B1	–	–	–	–	–	–		

Variety	Belgium	Denmark	France	Greece	Ireland	Italy	Netherlands	United Kingdom	FR Germany	Descriptive reference	Synonym
Guardian	–	–	–	–	–	–	–	A2	–	14,17	Alnwich Castle
Gunda	–	–	A	–	–	–	–	–	–	19,11	
Gustav Adolf	–	B1	–	–	–	–	–	–	–	18	
Gusto	–	–	A	–	–	–	–	–	A2	18,11,21	Grazia
Haddingtons	–	–	–	–	B2	–	–	–	–	8	
Haig	–	–	–	–	–	–	B	–	–	11	
Hanne	–	–	–	–	–	–	–	–	D1	18,11	
Hansa	A	B1,2,C3	A	–	B1	–	A	–	B2,D1	18,19,1,11	
Harbinger	–	–	–	–	–	–	–	A1	–	16,17,8	
Harmony	–	–	–	–	–	–	–	B1	–	14	
Hassia	–	–	–	–	–	–	–	B1	–	11	
Heather	A,C	–	–	–	–	–	–	A1,C1	–		
Heideniere	–	–	A	–	–	–	–	–	–	11,21	
Heideperle	–	–	–	–	–	–	–	–	D1	11	

Variety											
Heiderot	–	–	–	–	–	–	–	–	D1	11	
Heidrun	–	B1	B	–	–	–	B	–	A2	18,11,21	
Hela	–	B1,C3	A	–	–	–	B	–	A2	18,19,1,11	
Hera	–	B1	–	–	–	–	B	B1	–	11	
Herakles	–	–	–	–	–	C	–	–	–		
Herald	–	–	–	–	B1,2	–	–	A1	D1	16,8,11	
Herbstrote	–	–	D	–	–	–	–	–	–	5	
Herd Laddie	–	–	–	–	–	–	–	A1	–	16,17,8	
Hermes	–	–	–	–	–	–	–	–	D1	11,21	
Hertha	–	B1	A	–	–	D	A	–	A2	11,12,18,21 31,10	
Herulia	–	–	D	–	–	–	–	–	–	6	
Hibernia	–	–	–	–	–	–	–	A1	–	8	
Hinbinskii Ranny	–	–	D	–	–	–	–	–	–	11	Hibinsky Ranny
Hibinsky Ranny	–	–	–	–	–	–	–	–	D1		Hinbinskii Ranny
Hilla	–	–	–	–	–	–	B	–	–	11	
Hindenburg	–	B1	D	–	–	–	A	–	–	11,17	
Hochprozentige	–	B1	–	–	–	–	–	–	–	11	
Hohe Allerfruheste Gelbe	–	–	A	–	–	–	B	–	–	11,21	

Variety	Belgium	Denmark	France	Greece	Ireland	Italy	Netherlands	United Kingdom	FR Germany	Descriptive reference	Synonym
Holde	A	–	A	–	–	–	A	–	–	18,11,21	
Hollande De Roscoff	–	–	D	–	–	–	–	–	–	5	
Home Guard	–	–	D	–	A1,2,C3	–	A	A1,2,3,5,– B4,C1	–	13,15,16,7	
Horizon	–	–	–	–	–	–	A	–	–	32	
Houma	–	–	D	–	–	–	–	B1	D1	23,11	
Huayro	–	–	C	–	–	–	–	–	–		
Hudson	–	–	–	–	–	–	A	–	–	11	
Huinkul	–	–	C	–	–	–	–	A2,C4	–	30	
Humalda	–	–	A	–	A1,2	–	A	B1	–	11,21	
Hunter	–	–	D	–	–	–	A	–	–	11,24	
Hunters Gold	–	–	–	–	–	–	–	A1	–		
Huron	–	–	D	–	–	–	–	–	–	23,11,24	
Hutten	–	–	D	–	–	–	–	–	–	29	
Hydra	–	–	A	–	–	–	A	B1	D1	19,11,21,31	

Ica Guantiva	–	–	C	–	–	–	–	–	–	30	
Ica Narino	–	–	C	–	–	–	–	–	–	30	
Ica Nevado	–	–	C	–	–	–	–	–	–	30	
Ica San Jorge	–	–	C	–	–	–	–	–	–	30	
Ida	–	–	–	–	–	D	–	–	A2	18,21	
Idaho	–	–	–	–	B2	–	–	–	–	11	
Ideaal	–	–	A	–	–	–	–	–	–	11	
Iduna	–	–	D	–	–	–	–	–	–	11	
Igor	–	–	–	–	B1	–	A	–	–	11,21	
IJsselster	–	–	D	–	B1	–	A	–	–	11,21	Ysselster
Ilona	A	–	A	–	–	–	A	–	A2	18,11,21	
Ilse	–	–	–	–	–	–	–	B1	A2	18,11,21	
Imandra	–	B1	D	–	–	–	B	–	–	11	
Imilla Blanca	–	–	C	–	–	–	–	–	–	30	
Immune Ashleaf	–	–	–	–	B2	–	–	A1	–	8,16,17	Juli
Immune Seedling	–	–	–	–	B2	–	–	–	–		
Immune Up to Date	–	–	–	–	B2	–	–	–	–		
Imperia	–	–	D	–	B1,2	–	–	–	–	5	
Incastar	–	–	–	–	–	–	–	A1,2	–	13,11	
Incomer	–	–	–	–	–	–	–	B1	–	8	

Variety	Belgium	Denmark	France	Greece	Ireland	Italy	Netherlands	United Kingdom	FR Germany	Descriptive reference	Synonym
Indira	–	–	–	–	–	–	–	–	A2	18,11,21	
Industrie	–	–	–	–	–	–	B	–	D1	11,17	
Industrie Framboise	–	–	C	–	–	–	–	–	–		
Inga	–	–	–	–	–	–	–	B1	–		
Inis	*See* Iris										Iris
Institut De Beauvais	–	–	A	–	–	–	B	–	–	4,11,21	
International Kidney	–	–	A,C	–	A2,B1	–	–	A1,3	–	11,13,16,17	Jersey Royal
Intock	–	–	–	–	–	–	–	A1	–	14	
Inverness Favourite	–	–	–	–	–	–	–	B1	–	8	
Invincible	–	–	–	–	–	–	–	B1	–	8	
Iodine Rogue	–	–	–	–	B2	–	–	–	–		
Irene	–	B1	A	–	A1,2,C3	–	A	B1,4	–	11,12,13,21	
Iris	–	–	A	–	–	C	A	B1	A2	18,11,21	Inis
Irish Chieftain	–	–	–	–	B2	–	–	A1	–	8,16,17	

										America
Irish Cobbler	–	D	–	B2	C	B	–	–	23,11,24	–
Irish Peace	–	–	–	A2,B1	–	–	A1,3,5, C1	–	7,11,13,21	–
Irish Queen	–	–	–	B1,2	–	–	A1	–	16,8	–
Irish Whites	–	–	–	B2	–	–	B1	–	8	–
Irmgard	–	A	–	–	–	–	–	B2,C1	18,19,11,21	–
Iroise	–	A	B	–	C	A	–	–	4	–
Isabell	–	A	–	–	–	A	–	–	18,11,21,32	–
Isis	–	–	–	–	–	–	B1	–	17	–
Iskra	–	–	–	–	–	B	–	–	11	–
Isna	–	A	–	–	–	–	–	A2	18,11,21	–
Isle	–	–	–	–	–	–	B1	–	11	–
Isola	–	A	–	B1	–	A	B1,4	A2	18,19,11,21	–
Istok	–	–	–	–	–	A	–	–	11	–
Istrinskij	–	D	–	–	–	–	–	–	11	–
Iturrieta Temprana	*See* Bintje.	–	–	–	–	–	–	–	–	–
Izstades	–	D	–	–	–	–	–	–	21	Bintje
Jacoba	–	–	–	B2	–	–	–	–	–	–
Jacova	B1,2,C3	–	–	–	–	–	–	–	–	–
Jady	–	–	–	–	–	A	–	–	32	–

Variety	Belgium	Denmark	France	Greece	Ireland	Italy	Netherlands	United Kingdom	FR Germany	Descriptive reference	Synonym
Jaerla	–	B1,2,C3	A	B	B1	D	A	A1,3,2	D1	10,11,12,1,4	
Jakobi	–	–	D	–	–	–	A	–	–	11	
Jam	–	–	–	–	–	–	B	–	–	11	
Janet	–	–	–	–	–	–	–	A1,3	–		
Janka	–	–	–	–	–	–	A	–	–	20,11,21	
Jarolavskij	–	–	D	–	–	–	–	–	–		
Jaune De Hollande	–	–	–	–	–	–	–	B1	–		
Jaune D'or	–	–	D	–	–	–	B	–	–	11	Fransen
Jaune Du Pays De L'Aveyron	–	–	D	–	–	–	–	–	–		
Jemseg	–	–	–	–	–	D	–	–	–		
Jernak	–	–	D	–	–	–	–	–	–		
Jersey Royal	*See* International Kidney										International Kidney
Jessica	–	–	A	–	–	–	A	–	A2	18,11,21,32	

Variety												Synonym
Jetta	–	–	A	–	–	–	–	A	–	A2	18,19,11,21	
Jewel	–	–	–	–	C3	–	–	–	A1,3,C1	–		
Joelle	B	–	–	–	–	–	–	–	–	–		
Johanna	–	–	D	–	–	–	–	–	–	D1	19	
John Bull	–	–	–	–	B2	–	–	–	B1	–	8,14,16	Glenalmond
Jonsok	–	–	–	–	–	C	–	–	–	–		
Jose	–	–	A	–	–	–	–	A	–	–	11,12,4,21	Wilja
Jossing	–	–	D	–	–	–	–	–	–	–	11,21	
Jubel	–	B1	–	–	B2	–	–	B	–	D1	5,6,11	
Judika	–	–	–	–	–	–	–	–	–	A2	18,11,21	
Judith	A,C	–	–	–	–	–	–	–	–	–		
Jugovostocnyj	–	–	D	–	–	–	–	–	–	–	27	Jugowos-toschny
Juli	–	B1,C3	–	–	–	–	–	–	–	–	3,5,11,17	Ledvinky, Immune
Juliniere	–	–	–	–	–	–	–	B	–	–		Ashleaf
Juliver	–	–	A	–	–	–	–	A	B1	A2	18,11,21,31,32	
Juno	–	–	–	–	–	–	–	–	–	A2	18,11,21	
Juzanin	–	–	D	–	–	–	–	–	–	–		

Variety	Belgium	Denmark	France	Greece	Ireland	Italy	Netherlands	United Kingdom	FR Germany	Descriptive reference	Synonym
Kaimes Beauty	–	–	–	–	–	–	–	B1	–	14,16	
Kalinka	A	–	–	–	–	–	–	–	–		
Kameraz	–	–	A	–	–	–	–	–	–	11	Kameraz No.1
Kameraz No. 1	–	–	–	–	–	–	A	–	–	27	Kameraz
Kaptah	–	B1,2,C3	A	–	–	C	A	–	D1	1,2,11,21	Kaptah Vandel
Kaptah Vandel	*See* Kaptah										Kaptah
Kaptanok	–	–	–	–	–	–	–	–	D1	22,4	Kaptah
Kardinal	–	–	D	–	–	–	–	–	–	11	
Kastor	–	–	A	–	–	–	A	–	D1	4,11,21	
Kathadin	–	B1	D	–	B1,2	–	A	A1,B2	–	23,11,14,21	Kathadin Hollandaise, Golden
Kathadin Hollandaise *See* Kathadin *and* Golden											
Katja	–	–	–	–	–	–	–	–	A2	18,11,21	Kathadin
Katie Glover	–	–	–	–	–	–	–	B1	–	17,8	

Variety												Notes
Katiusa	–	–	D	–	–	–	–	–	–	–	27	Katjuscha
Kavkazskij	–	–	D	–	–	–	–	–	–	–	27	Kawkaski
Keltia	–	–	A	–	–	–	–	A	B1	–	4,11,21	
Kenavo	–	–	D	–	–	–	–	–	–	–	11	
Kennebec	–	B1,C3	D	B	A2,B1,C3	–	C	A	A1,2,3,5,C1	D1	23,1,10,11,12	
Kenta	–	–	D	–	–	–	–	–	–	–	28	
Kenva	–	B1,2,C3	–	–	–	–	–	A	–	D1	1,2,11,21	
Kenzy	–	–	–	–	–	–	–	–	A3,B1,C1	–		
Kepplestone Kidney	–	–	D	–	–	–	–	–	A1	–	8,16	
Kerebel	–	–	D	–	–	–	–	–	B1	–	11	
Kermor	–	–	A	–	–	–	C	A	–	–	4	
Kerne	–	–	A	–	–	–	–	A	–	–	11,14	
Kero	–	–	–	–	–	–	–	–	–	A2	18,11,21	
Kerpondy	–	–	A,C	–	–	–	–	A	B1	–	4,11,14,19	
Kerr's Pink	–	B1	D	–	A1,2,C3	–	–	–	A1,2,3,5,–B4,C1	–	13,15,16,3,5	Green Budded Kerr's Pink Rogue
Kerr's Pink (Green Bud)	–	–	–	–	B2	–	–	–	–	–	8	

Variety	Belgium	Denmark	France	Greece	Ireland	Italy	Netherlands	United Kingdom	FR Germany	Descriptive reference	Synonym
Kerr's Pink Purple Flower	–	–	–	–	B2	–	–	–	–		
Kerr's Pink White	–	–	–	–	B2	–	–	–	–	14	
Keswick	–	–	–	–	–	–	–	B1	–	23,11,24	
King Edward	–	B1,C3	D	–	A1,2,C3	–	A	A1,2,3,5 C1,4	–	13,15,16,1,3 5,11,21,22	King Edward VII
King Edward VII		*See King Edward*								17,7,8	King Edward
King Edward (Wilding)	–	–	–	–	B2	–	–	–	–		
King George	–	–	–	–	B2	–	–	A1,3	–	13,16,11,21,8, 17	
Kingston	–	–	–	–	A2,B1	–	A	A1,2,3,5 C1	–	13,11,21	
Kirsty	–	–	–	–	A2,B1	–	–	A1,3,5, C1,4,C4	–	13	
Kismet	–	–	–	–	–	–	–	B1,C1	–		
Kisvardai Rozsa	–	–	–	–	–	–	–	–	D1	11	
Klawina	–	–	–	–	–	–	A	–	–	31,32	

Variety									D1		
Kleinod	–	–	–	–	–	–	–	–	–	–	
Klio	–	–	D	–	–	–	–	–	–		
Klondyke	–	–	–	–	A2	–	–	A1,2,3,5	–	11,13,21	
Kmiec	–	–	D	–	–	–	–	–	–	28	
K of K	–	–	–	–	B2	–	–	A1	–	17,8	
Komeet	–	–	D	–	–	–	–	–	–	11	
Kondor	–	–	–	–	–	C	A	A1,2,3,5 C1,4	–	10,12,32	
Koning	–	–	A	–	–	–	A	A1	–	11,21	
Konsuragis	–	–	–	–	–	–	B	–	–	11	
Koopmans Blauwe	–	B1	D	–	–	–	A	–	–	11	
Korensevsky	–	–	–	–	–	–	A	–	–	11	Koronevsky
Korrigane	–	–	A	B	–	C	A	–	–	4	
Kotnov	–	–	D	–	–	–	–	–	–	11	
Krasava	–	–	A	–	–	–	A	–	–	4,11,21	
Krebsfeste Kaiserkrone	–	–	D	–	–	–	–	–	–	29	
Kristalla	–	–	–	–	–	–	–	–	A2	18,11,21	
Krodal	–	–	–	–	–	–	A	–	–	31,32	
Krodana	–	–	–	–	–	–	A	–	–	32	

Variety	Belgium	Denmark	France	Greece	Ireland	Italy	Netherlands	United Kingdom	FR Germany	Descriptive reference	Synonym
Krodeka	–	–	–	–	–	–	A	–	–	32	
Kroline	–	–	–	–	–	–	A	–	–	31,32	
Krolisa	–	B1	–	–	–	–	A	B1	–	31,32	
Kromargretha	–	–	A	–	–	–	A	A2,B1	–	31,32	
Kromaris	–	–	–	–	–	–	A	–	–	32	
Kronia	–	–	A	–	–	–	A	–	–	11,31	
Krostar	–	–	A	–	–	–	A	–	D1	10,11,12,21	Krostaria
Krostaria	*See Krostar*									4	Krostar
Kroto	–	–	A	–	–	–	–	–	–	31,32	
Kurrell	–	–	–	–	–	–	–	A1	–	11	
Kvetuse	–	–	D	–	–	–	–	–	–	6	
La Chipper	–	–	D	–	–	–	B	–	–	7,11	
Laila	–	–	–	–	–	C	–	–	–		
Lajana	–	–	–	–	–	–	–	B1,C1	A2	18,11,21	

Variety											
Lama	–	–	–	–	–	–	–	–	D1	11,19,21	–
Lamia	–	–	A	B	–	C	A	B1,C1	–	4,11,32	–
Langworthy	–	–	–	–	B2	–	–	B1	–	17	Maincrop
La Rouge	A	–	D	–	–	–	B	–	–	11	–
Laura	–	–	–	–	–	–	–	–	D1	11,21	–
La Verta	–	–	D	–	–	–	–	–	–	–	–
Leander	–	–	A	–	–	–	A	–	–	4,11,21	–
Leda	–	–	–	–	–	C	–	–	–	–	–
Ledvinky	–	–	C	–	–	–	–	–	–	5	Juli
Lekkerlander	–	–	A	–	–	–	A	–	D1	11,12,21	–
Lenape	–	–	D	–	–	–	B	–	–	11	–
Lenino	*See* Capella	–	–	–	–	C	–	–	–	–	Capella
Lenon	–	–	–	–	–	–	–	–	–	–	–
Lenor	–	–	A	–	–	–	A	–	–	4,11	–
Lentia	–	–	D	–	–	–	B	–	–	11,21	Linzer Starke
Leo	–	–	A	–	–	–	–	–	D1	19,11	–
Leona	–	–	D	–	–	–	–	–	–	11	–
Lerche	–	–	D	–	–	–	–	–	D1	19,11	–
Lewis Black	–	–	–	–	–	–	–	A1	–	–	–

Variety	Belgium	Denmark	France	Greece	Ireland	Italy	Netherlands	United Kingdom	FR Germany	Descriptive reference	Synonym
Libertas	–	B1	D	–	B1,2	–	A	A2,3,B1, 4	A2,3,B1, –	11,7,14	
Libelle	–	–	–	–	–	–	–	C4	–		
Lichie Industrie	–	–	–	–	–	–	B	–	–		
Lichte Rode Star	–	–	–	–	–	–	B	–	–		
Lido	–	–	–	–	–	–	–	B1	–		
Linda	–	–	B	–	–	–	–	–	A2	18,11,21	
Linzer Delikatess	–	–	–	–	–	C	–	–	D1	11,21	
Linzer Gelbe	–	–	–	–	–	C	–	–	D1	11,21	
Linzer Rose	–	–	D	–	–	C	A	–	D1	11,21,32	
Linzer Starke	–	–	–	–	–	–	–	–	D1	11,21	Lentia
Liro	–	–	–	–	–	–	–	A1,3	–	11,13,21	
Lisa	–	–	–	–	–	C	–	B1	D1	11,21	
Listowel Kerry Rogue	–	–	–	–	B2	–	–	–	–		
Liubimec (1)	–	–	C	–	–	–	–	–	–	11	
Lizen	–	–	A,C	B	–	C	A	–	–	4,11,21	

Variety										
Lochar	–	–	–	B2	–	–	B1	–	16,17,8	
Lola	–	A	B	C3	D	A	A1,3,C1	–	4,11,21,32	Sjitskii
Long Island White	–	–	–	B2	–	–	–	–	–	
Long Leaf	–	–	–	–	–	–	B1	–	11	
Lorch	–	–	–	–	–	A	–	–	8	
Lord Rosebery	–	–	–	–	–	–	A1	–	–	
Lord Scone	–	–	–	–	–	–	B1	–	–	
Lori	–	A	–	–	–	A	–	D1	11,19	
Losicki	–	D	–	–	–	A	–	–	11	
Lumar	–	–	–	–	–	A	–	–	31,32	
Lumper	–	–	–	–	–	–	A1	–	8	
Lusewitzer	–	–	–	–	–	–	–	D1	–	
Lutetia	–	–	B	–	D	–	B1	–	–	
Lutina	–	–	–	–	D	A	–	–	32	
Lux	–	A	–	–	–	–	–	–	11	
Luxor	–	A	–	–	–	–	–	–	–	
L'vovskij Belyj	–	D	–	–	–	–	–	–	–	
Lymn Grey	–	–	–	B2	–	–	A1	–	8,16	

Variety	Belgium	Denmark	France	Greece	Ireland	Italy	Netherlands	United Kingdom	FR Germany	Descriptive reference	Synonym
Madam	A	–	A	–	–	–	A	A1,3,C1	–	11,21	
Mador	–	–	A	–	–	–	–	–	–	11	
Magali	–	–	A	–	–	–	–	–	–	11	
Magna	–	–	D	–	–	–	–	–	–	11	
Magneto	–	–	D	–	–	–	A	–	–	11	
Magnificent	–	–	–	–	–	–	–	B1	–	16	
Magnum Bonum	–	B1	–	–	B1	–	B	A1	–	14,16,17,8,11	
Magura	–	B1	D	–	–	–	A	–	–	11,21	
Maincrop	*See* Langworthy										Langworthy
Maja	–	–	A	–	–	–	A	–	A2	18,11,21,31	
Majestic	–	B1,C3	A	–	A1,2	C	A	A1,2,3,5,B4,C1	–	13,15,16,17,11	
Malika	–	–	A	–	–	–	A	–	–	4,11,21,32	
Mandel	–	–	–	–	–	–	A	–	–	11	
Mandy	–	–	–	–	–	–	–	–	A2	18,11,21	

Variety											
Manna	—	—	—	—	A2,B1	C	A	A1,5,3, C1	—	10,11,12,4	
Mansholt 67–295(Arsy)	—	—	A	—	B1	—	A	—	—	32	Arsy
Mansour	—	—	—	—	—	C	A	B1,C1	—	31,32	
Mansur	—	—	—	B	—	—	—	—	—		
Manuela	—	—	—	—	—	—	A	—	—	11,21	
Mara	—	—	A	—	—	—	A	—	—	32,11,21	
Marava	—	B1,2,C3	—	—	—	—	—	—	—	2	
Marco	—	—	—	—	—	—	—	A2,D1	—	18,11,21	
Marconi	—	—	—	—	—	—	—	B1	—	14	
Marfona	—	—	A	B	—	C	A	A1,2,3,5 C1	—	10,11,12,4	
Maria	—	—	C	—	—	—	—	B1	—	11,21	
Mariana	—	—	A	B	—	C	A	—	—	4,32	
Marina	—	—	—	—	—	—	—	B1	—		
Mariella	—	B1	A	—	—	—	—	—	—	11,21	
Marijke	—	B1	—	—	A1,2	C	A	—	—	10,11,12,13 31	Maryke
Mariline	A	—	B,D	—	—	—	A	—	—	11,21	
Marion	A	B1,C3	A	—	—	—	A	A2,D1	—	18,1,11,21,31	

Variety	Belgium	Denmark	France	Greece	Ireland	Italy	Netherlands	United Kingdom	FR Germany	Descriptive reference	Synonym
Maris Anchor	–	–	A	–	A1,2	–	A	A1,5	–	13,11,21	
Maris Bard	–	–	A	–	A1,2	C	A	A1,2,3,5 C1	–	13,15,11,21	
Maris Page	–	–	A	–	A1,2	–	–	A1,2,3	–	13,11,21	
Maris Peer	–	–	–	–	A1,2,C3	–	A	A1,2,3,5 B4,C1	–	13,15,7,11	
Maris Piper	–	B1	A	–	A1,2,C3	C	A	A1,2,3,5 B4,C1	–	13,15,11,21	
Maritta	–	B1	A	–	B1,2	–	A	B1,4	B2,D1	18,19,4,11	
Marjolin	–	–	A	–	–	–	–	–	–	5	
Marlene	–	–	A	–	–	–	A	–	–	4,11,21	
Marna	–	–	–	–	C3	–	–	–	–		
Marne	–	–	–	–	–	–	A	–	–	31,32	
Marquis of Bute	–	–	–	–	B2	–	–	B1	–	16,17,8	
Marygold	–	–	D	–	–	–	–	–	–	11	
Maryke	–	–	A	–	A1,2	–	–	–	D1	4,13,19,21	Marijke

Variety											
Matador	–	D	–	–	–	–	–	–	–	11	
Matvecvskij	–	D	–	–	–	–	–	–	–		
Maud Meg	–	–	–	–	–	–	B1	–	–		
Mauve Queen	–	–	–	–	–	–	B1	–	–	14	
Max Delbruck	–	D	–	–	–	–	–	–	–	5	
May Queen	–	D	–	B2,C3	–	B	B1	–	–	13,16,17,5, 7,11	
McBeths Castle	–	–	–	–	–	–	A1	–	–		
Meerlander	–	–	–	–	–	A	–	–	–	11,12,21	
Meerster	–	–	–	B1,2	–	–	–	–	–	11	
Meins Early Round	–	–	–	B2	–	–	A1	–	–	8,16	
Meise	–	D	–	–	–	B	–	–	–	11	
Meliora	–	A	–	–	–	A	–	–	–	11,4,21	
Menominee	B1	–	–	–	–	–	–	–	–	23,11	
Mensa	–	C,D	–	–	–	–	–	–	D1	19,11	
Mentor	B1	A	–	A2,B1	–	A	–	–	B2,D1	10,11,12,7	
Merkur	–	D	–	–	C	–	–	–	C1	19,11,21	Alava
Merrimack	–	B	–	–	–	B	–	–	–	23,11	
Mesaba	–	D	–	–	–	–	–	–	–	23,11	
Midlothian Early	*See* Duke of York, Eersteling, Eerstelingen *and* Erstling										Duke of York

Variety	Belgium	Denmark	France	Greece	Ireland	Italy	Netherlands	United Kingdom	FR Germany	Descriptive reference	Synonym
Mighty Atom	–	–	–	–	–	–	–	A1	–		
Mila	–	–	–	–	–	D	–	–	–		
Millars Beauty	–	–	–	–	B2	–	–	–	–	8	
Mill St Hero	–	–	–	–	B2	–	–	–	–	8	
Milva	–	B1	–	–	–	–	–	–	–	2,11	
Minea	–	B1,2,C3	–	–	–	–	A	A1	D1	1,2,11,13,21	Minea Vandel
Minerva	–	–	–	B	–	D	A	–	–	32	
Minke	–	–	–	–	–	–	A	–	–	31	
Minsand	–	B1	–	–	–	–	–	A1,3	–	2,11,13,21	
Mira	–	–	D	–	–	–	–	–	D1	11,21	Ora
Miranda	–	B1	A	–	–	–	A	–	A1,2	18,1,11,21,32	
Mireille	–	–	A	–	–	–	–	–	–	11	
Mirka	A	B1	A	–	–	–	A	B1,4	–	10,11,12,21	
Mirton Pearl	–	–	–	B	–	–	–	–	–		

Variety											Synonym
Mistral	–	–	–	–	D	–	–	–	–	–	
Mittelfruhe	B1	D	–	–	–	A	–	–	–	19,3,11	Bohms Mi Melfruhe
Mizen	–	–	B	A1,2,C3	–	–	A1,3	–	–	9,11,13,21	
Mohawk	–	D	–	–	–	–	–	–	–	23,22	
Moira	–	–	–	–	–	–	A1,2,3,5,C1	–	–	23,22	
Monalisa	–	A	B	C3	C	A	A1,3,C1	–	–	10,11,12,4	
Moni	–	A	–	–	–	–	B1	A2	–	18,11,21	
Monika	–	D	–	B1,2	–	–	B1	–	–	11	
Monitor	–	A	–	–	–	A	B1	–	–	11,4,21	
Monocraat	–	D	–	–	–	–	–	–	–	11	
Monona	–	–	–	–	–	A	B1	D1	–	11	
Moncerrate	–	–	–	–	–	–	A2	–	–		
Montana	–	D	B	–	–	B	A1,2,3	–	–	11,13,21	
Monza	–	–	B	–	–	–	–	A2	–	18,11,21	
Moray	–	–	–	–	–	–	A1,C1	–	–		
Morene	–	A	B	–	C	A	A1,3,C1	–	–	10,31,32,12	
Morgane	–	A	–	–	–	–	–	–	–	11	
Move	*See Mowe*	–	–	–	–	–	–	–	–		Mowe
Mowe	–	D	–	–	–	–	–	D1	–	19,11	Move

Variety	Belgium	Denmark	France	Greece	Ireland	Italy	Netherlands	United Kingdom	FR Germany	Descriptive reference	Synonym
Mr. Bresee	–	–	–	–	–	–	–	A1	–	8,17	
Multa	–	–	A	–	A1,2	–	A	A2,B1	–	10,11,12,4	
Muncel	–	–	–	–	–	C	–	–	–		
Munstersen	–	–	–	–	–	–	B	–	–	11	
Muntinga (17)	–	–	D	–	–	–	–	–	–	11	
Mural	–	–	–	–	–	–	A	–	–	32	
Murillo	–	–	–	–	–	C	A	–	–	11,32	
Murmanskij	–	–	D	–	–	–	–	–	–	27	Murmanski
Musana	–	–	–	–	–	–	–	–	D1		
Myatts Ashleaf	–	–	–	–	–	–	–	A1,2	–	8,16,17	Ashleaf Kidney
Nadia	A	B1	A,C	–	–	–	–	B1,C1	–	11	
Nahodka	–	–	D	–	–	–	–	–	–	27	Nashodka
Nascor	–	–	–	–	A1,2	–	A	–	–	11	
Natalie	–	B1	–	–	–	–	–	–	A2	18,21	

Variety										Prenor (F)
Nederlander	–	–	D	–	–	–	–	–	–	11
Nemae	–	–	A	–	–	–	–	–	–	19,11
Nervia	A	–	A	–	–	–	–	–	–	11
Netted Gem	–	–	A	–	B2	–	–	–	–	11,24
Nicola	–	B1.C3	A	–	B1	C	A	B1	A2	18,1,4,10,11, 12,31.
Ninetyfold	–	B1	D	–	B2	–	B	A1,3	–	13,16,17,8,11
Nippigon	–	–	–	B	–	C	–	–	–	
Nithsdale	–	–	–	–	–	–	–	B1	–	17,8
Noella	–	–	–	–	–	–	A	–	–	32
Noordeling	–	–	D	–	–	–	A	–	–	11
Noordstar	–	B1	D	–	–	–	–	–	–	11
Nora	–	–	–	–	–	–	A	–	–	11,21
Norchip	A	–	A	–	A1,2	C	A	–	–	11,21,24
Nordlicht	–	–	–	–	–	C	–	–	A2	18,11,21
Nord 23K33	–	–	–	–	–	–	A	–	–	
Nordstern	–	–	A	–	–	–	A	–	A2	18,11,21
Norkota	–	–	D	–	–	–	–	–	–	23,11
Norland	B	–	D	–	–	C	A	–	–	23,11,24
Northern B	–	–	–	–	–	–	–	B1	–	16

Variety	Belgium	Denmark	France	Greece	Ireland	Italy	Netherlands	United Kingdom	FR Germany	Descriptive reference	Synonym
Northern Star	–	–	–	–	B2	–	–	B1	–	16,17,8	
North Island Skerry	–	–	–	–	B2	–	–	–	–		
Nova	–	–	A	–	–	–	A	–	–	11	
N.S.O.Potato	–	–	D	–	–	–	–	–	–		
Oak Park Amber	*See* Amber									21	Amber
Oak Park Avenger	*See* Corrib									21	Corrib
Oak Park Beauty	*See* Cara									11,21	Cara
Oak Park Bounty	*See* Clada									21	Clada
Oberarnbacher Fruhe	–	–	A	–	–	–	A	–	–	19,11	
Oceane	–	–	–	–	–	C	–	–	–		
Octavia	–	B1,2,C3	B	–	–	–	A	–	D1	1,11,21,31	
Oda	–	–	–	–	–	–	–	–	D1	19,11	
Odenwalder Blaue	–	–	D	–	–	–	–	–	D1	11	
O'Farrells Seedling	–	–	–	–	B2	–	–	–	–		

Variety											
Ogenek (Ogonjok)	–	–	D	–	–	–	A	–	–	11	
Oktkabrenok	–	–	–	–	–	–	B	–	–	11	
Olalla	–	–	–	–	–	–	A	–	–	11,21	
Old Black	–	–	–	–	–	–	–	A1	–		
Old Lumper	–	–	–	–	B2	–	–	–	–	18	
Olinda	–	–	A	–	–	C	A	B1,C1	–	10,11,21	
Olympia	–	B1	D	–	–	–	–	–	C1	19,11,21	
Ombra	–	–	–	–	–	–	A	A1	–	11,21	
Omega	–	–	A	–	–	–	A	–	D1	11,21,31,32	
Ona	–	–	–	–	A2,B1	–	–	–	–	11,21	
Ontario	–	B1	–	–	B1	–	B	B1	–	23,11	
Opperdoese Ronde	–	–	–	–	–	–	B	–	–	11	Mira
Ora	–	–	A	–	A1,2	–	A	–	–	11,21	
Origo	–	–	–	B	–	C	A	–	–	32	
Orion	–	B1	–	–	A1,2	–	A	A1,3	–	13,16,11,21	
Orion (NL)	–	–	–	–	–	–	–	B1	–		
Oromonte	–	–	–	B	–	–	–	–	D1		
O'Sirene	–	–	–	B	–	C	–	–	–		
Ornament	–	–	–	–	–	–	–	A1,3	–		

Variety	Belgium	Denmark	France	Greece	Ireland	Italy	Netherlands	United Kingdom	FR Germany	Descriptive reference	Synonym
Oslava	*See* Oslava Tcheque										Oslava Tcheque
Oslava Tcheque	–	–	D	–	–	–	–	–	–	28	Oslava
Ostara	–	B1,C3	A,C	–	B1	–	A	–	A2	10,11,12,1,4	
Ostbote	–	–	A	–	–	–	A	–	D1	11	
Ostenbrink	–	–	–	–	A1,2	–	–	–	–		
Ovalgelbe	–	B1	D	–	–	–	–	–	–	3,5	
Oxford Early	–	–	–	–	B2	–	–	–	–	8	
Paarsput	–	–	–	–	–	–	B	–	–	11	
Pallas	–	–	–	–	–	–	A	–	–	11	
Palma	–	–	A	B	–	–	A	–	A2	18,1,4,11,21,32	
Pamir	–	–	–	–	–	–	B	–	–	19,11	
Pana	–	B1	A	–	–	–	A	–	–	19,11	
Pandora	–	B1	A	–	–	–	A	–	–	11	

President

Variety											
Pansta	–	B1	A	–	–	–	–	A	A2,B1	–	11,21,31
Panther	–	–	D	–	–	–	–	B	–	D1	19,4,11,21
Pantucha	–	–	–	–	–	–	–	B	–	–	11
Paragon	–	–	–	–	–	–	–	–	A1,2	–	11,13,21
Parel	–	B1	–	–	–	–	C	A	–	–	11,12,21
Parnassia	–	–	–	–	B1,2	–	–	A	B1	D1	19,11,21
Passat	–	–	–	–	–	–	–	A	–	–	11
Patawi	–	–	A	–	–	–	–	A	–	–	11
Patricia	A	–	–	–	–	–	–	–	–	–	
Patrones	–	B1,C3	A	–	A1,2	–	–	A	A1,2,5	–	10,11,12,1,7
Paula	–	–	–	–	–	–	–	–	–	D1	
Paul Kruger	–	–	–	–	–	–	–	A	–	–	11
Pavo	–	–	–	–	–	–	–	–	–	D1	19,11
Pawnee	–	–	–	–	–	–	–	–	B1	–	23
Peach Bloom	–	–	–	–	B2	–	–	–	A1	–	8,16
Peconic	–	–	–	–	–	–	–	A	–	–	11
Peerless	–	–	–	–	B2	–	–	–	B1	–	23,8
Penobscot	–	–	–	–	–	–	–	B	–	–	
Penta	–	–	–	–	–	–	–	–	A1,3,C1	–	
Pentland Ace	–	–	–	–	A1,2	–	–	–	A1,2,3,4	–	16,14,11

Variety	Belgium	Denmark	France	Greece	Ireland	Italy	Netherlands	United Kingdom	FR Germany	Descriptive reference	Synonym
Pentland Beauty	–	–	A	–	A1,2	–	–	A1,2	–	13,16,7,11	
Pentland Crown	–	–	A	–	A1,2,C3	–	A	A1,2,3,5,–B4,C1	A1,2,3,5,–	13,15,16,7,11	
Pentland Dell	–	–	A	–	A1,2,C3	–	A	A1,3,5,B4,C1	A1,3,5,B4,C1	13,15,7,11	
Pentland Envoy	–	–	–	–	A1	–	A	A1,2,B4	A1,2,B4	13,11	
Pentland Falcon	–	–	–	–	B1	–	A	A1,2,B4	A1,2,B4	13,11	
Pentland Glory	–	–	–	–	A2,B1	–	–	A1,2	–	13,7,11	
Pentland Hawk	–	–	–	–	A1,2,C3	–	A	A1,2,3,5,B4,C1	A1,2,3,5,–	13,15,11,21	
Pentland Ivory	–	–	A	–	A1,2,C3	–	A	A1,2,3,5,C1	A1,2,3,5,–	13,15,11,21	
Pentland Javelin	–	–	A	–	A1,2	–	B	A1,2,3,5,C1	A1,2,3,5,–	13,15,11,21	
Pentland Kappa	–	–	–	–	–	–	–	A1	–	14	
Pentland Lustre	–	–	–	–	A1,2	–	B	A1,2,3,C1	–	13,15,11,21	

Variety										Pinaza
Pentland Marble	—	—	A	—	—	A	A1,2	—	13,11,21	
Pentland Meteor	—	—	—	B1	—	—	A1,2	—	13,11,21	
Pentland Raven	—	—	A	—	—	—	A1,2	—	13,11,21	
Pentland Squire	—	—	A	A1,2,C3	—	—	A1,2,3,5,—C1	—	13,15,11,21	
Pepita	A	—	A	—	C	B	A	—	4,11,21,32	
Pepo	—	—	D	B1	—	—	A1,B2	—	11	
Percoz	—	—	—	—	C	—	—	—		
Perle Rose	—	—	A	—	—	—	—	—	11	
Persenk	A	—	—	—	—	—	—	—		
Perth Favourite	—	—	—	—	—	—	B1	—		
Petra	—	—	A	—	—	A	—	—	32	
Peyma	—	—	D	—	—	—	—	—		
Pierwiosnek	—	—	D	—	—	B	—	—	20,11,21	
Pimpernel	—	B1	D	A2,B1,C3	—	A	A2,3,B1	—	11,12,14,21	
Pinaza S. *juzepciukii*	—	—	C	—	—	—	—	—	30	
Pink Duke of York	—	—	—	—	—	—	A1,C1	—		
Pink Fir Apple	—	—	A	—	—	—	A1,2,3,C1	—	13,21	
Pinki	—	—	—	—	—	—	B1,C1	A2	18,11,21	

Variety	Belgium	Denmark	France	Greece	Ireland	Italy	Netherlands	United Kingdom	FR Germany	Descriptive reference	Synonym
Pionier	–	–	A	–	A1,2	–	A	–	–	11	
Pirat Rose	–	–	–	–	B1	–	–	–	–		
Pirmunes	–	–	D	–	–	–	–	–	–	11	
Pirola	–	B1	A	–	–	–	A	–	–	18,11,21,31,32	
Pito	–	–	–	–	–	–	A	–	–	11,21	
Planeta	–	–	–	–	–	–	A	–	–		
Plymouth	–	–	D	–	–	–	–	–	–	23	
Pocamas	–	B1	–	–	–	–	–	–	–	11	
Podzola	A	–	A	–	–	C	A	A1	–	31,32	
Poet	–	–	D	–	–	–	–	–	–	11	
Pollock's Pink Early	–	–	–	–	–	–	–	A1	–	11,14	
Pommerant	–	B1	A	–	A1,2	–	A	B1	–	11,21	
Pompadour	–	B1	A	–	–	–	A	B1	–	11,21	
Pontiac	–	–	D	–	–	–	–	–	–	23,11	

Populair	–	–	–	–	B1	–	A	–	–	11	–
Porta	–	–	–	–	–	–	B	–	D1	19,11,21	–
Posmo	–	B1,2,C3	A	–	–	–	–	–	–	1,4,11,21	–
Prefect	–	B1	D	–	–	–	A	–	–	11	–
Prelanda	–	–	A	–	–	–	–	–	–	11	–
Premiere	–	–	A	B	–	C	A	A1,2,3,- C1	–	10,11,12,4,31	Paul Kruger
Prenor	–	–	A	–	–	–	–	–	–	4	–
Present	–	–	D	–	–	–	–	–	–	11	–
President	–	–	–	–	A1,2	–	–	A1	–	11,8,14,16	–
Prestkvern	–	–	D	–	–	–	–	–	–	11	–
Preuszen	–	–	–	–	–	–	B	–	–	–	–
Prevalent	A	B1	A	–	A2,B1	–	A	A	–	10,11,12,4,31	–
Preziosa	–	–	A	–	–	–	–	–	–	11	–
Pride of Bute	–	–	–	–	B2	–	–	A1	–	8,16	–
Pride of Perth	–	–	–	–	B2	–	–	–	–	–	–
Prickulskij	–	–	D	–	–	–	–	–	–	27	–
Prickulskij Ranny	–	B1	–	–	–	–	B	–	–	11	–
Prima	A	–	A	–	–	C	A	–	A2	18,19,4,11	–
Primabel	–	–	A	–	–	–	–	–	–	11	–

Variety	Belgium	Denmark	France	Greece	Ireland	Italy	Netherlands	United Kingdom	FR Germany	Descriptive reference	Synonym
Primerose	–	–	A	–	–	–	A	–	–	11	
Primor	–	–	–	–	–	C	–	–	–		
Primula	–	B1,2,C3	D	–	–	–	A	–	–	1,11,21,22	
Primura	–	–	A	–	A1,2	C	A	B1	–	10,11,12,4	
Prinslander	–	–	–	–	–	–	A	–	–	11	
Prinzess	–	–	D	–	–	–	A	–	D	18,11,21	
Prior	–	–	–	–	–	–	–	–	–	31,32	
Prisca	–	B1	D	–	–	–	–	–	–	6,27	
Priska	–	–	–	–	–	–	–	–	D1	19,11	
Priwal	–	–	A	–	–	–	A	–	–	11,4,21	
Prizetaker	–	–	–	–	–	–	–	B1	–		
Probaat	–	–	A	–	–	–	A	–	–	11,31,32	
Procura	–	B1,C3	A	–	–	–	A	B1	D1	10,11,12,1,4,31	
Producent	–	–	–	–	–	–	A	–	–	12	

White
Beauty of
Hebron

Prof. Broekema	–	B1	–	–	A1.2	–	–	B1	–	11,14
Prof. Duboys	–	–	D	–	–	–	–	–	–	–
Prof. Wohltmann	–	–	–	–	–	–	B	–	D1	11
Profijt	–	–	D	–	–	–	A	–	–	11
Promesse	–	C3	–	–	–	–	A	B1	–	10,12,31
Prominent	–	B1	A	–	–	–	A	A2.B1	–	10,11,12,4,31
Prosna	–	–	–	–	–	–	A	–	–	20,11,21
Proton	–	–	A.C	–	–	–	A	A2.B1	D1	11,31
Provita	–	–	A	–	A1.2	C	A	–	–	11,12,1,21,31
Provost	–	–	–	–	A2.B1	–	–	A1.2,3,5,-C1	–	13,11,21
Pruceres	–	B1	A	–	–	–	A	–	A2	11,18,21,31
Prudal	–	–	A	–	–	–	A	–	–	11
Prumex	–	–	A	–	–	–	A	–	–	11,21,31
Prymas	–	–	D	–	–	–	–	–	–	–
Pungo	–	–	D	–	–	–	B	–	–	23,11,14
Puntila	–	–	A	B	–	–	A	–	A2	18,11,21
Puritan	–	–	–	–	B2	–	–	A1	–	8,16

Variety	Belgium	Denmark	France	Greece	Ireland	Italy	Netherlands	United Kingdom	FR Germany	Descriptive reference	Synonym
Purple Champion	–	–	–	–	B2	–	–	–	–		
Puskinskij	–	–	D	–	–	–	–	–	–		
Quarta	–	B1	–	–	–	–	A	–	A2	18,11,21,32	
Queen Mary	*See* Royal Kidney										Royal Kidney
Queen of the South	–	–	–	–	–	–	–	B1	–		
Radosa	A	–	A	–	–	D	A	–	–	10,11,12,4	
Raeburns Gregor Cups	–	–	–	–	–	–	–	B1	–	8,16	
Ragna	–	B1	–	–	–	–	A	–	A2	18,11,21	
Rajka	–	B1	–	–	–	–	–	–	–	11	
Ranfurly Red	–	–	–	–	–	–	–	B1	–	8	
Ranger	–	–	–	–	–	–	–	B1	–	14	

Variety										
Rannij Zelotyj	11	–	–	–	–	–	–	D	–	
Raritan	11,24	–	–	A	C	–	B	–	–	
Ratte	4,11,21	–	–	A	–	–	–	A	–	Asparges
Razvaristii	11	–	–	B	–	–	–	–	–	
Reaal	11	–	B1	–	–	B1	–	–	–	
Reading Russet	16,17,8	–	A1	–	–	B2	–	–	–	
Realta	11	–	–	–	–	A1,B2	–	–	–	
Reanne	–	–	–	A	–	–	–	–	–	
Rebus	–	–	–	–	–	–	–	A	–	
Recent	32	–	–	A	D	–	–	–	–	
Record	11,7,13,15	–	A1,2,3,5,B4,C1,4	A	–	A1,2,C3	–	A	B1,C3	Record (S.M.)
Record (S.M.)	–	–	–	–	–	–	–	–	*See* Record	Record
Rector	11,14,17,4,31	–	–	A	–	–	–	A	–	
Reda	–	–	–	–	C	–	–	–	–	
Red Ashleaf	–	–	–	–	–	–	–	A	–	
Redbad	11,4,21	–	B1	A	–	–	–	A	–	
Red Cara	9,11,21	–	A1,3,C1	A	–	A1,2,C3	–	–	–	
Red Craigs	–	–	–	–	–	–	–	A	–	
Red Craigs Royal	13,15,7,11	–	A1,2,3,C1	A	–	A1,2,C1	–	A	–	

Variety	Belgium	Denmark	France	Greece	Ireland	Italy	Netherlands	United Kingdom	FR Germany	Descriptive reference	Synonym
Red Cups	–	–	–	–	B2	–	–	–	–	8	
Red Drayton	–	–	–	–	–	–	–	A1,C1	–	–	
Red Fife	–	–	–	–	–	–	–	B1	–	14	
Red Kidney	–	–	–	–	B2	–	–	B1	–	8	
Red King	–	–	–	–	B2	–	–	–	–	–	
Red King Edward	–	–	–	–	–	–	–	A1,3,C1	–	13,15,11,21	
Red Lasoda	–	–	–	–	D3	–	B	–	–	23,11	
Red Letter	–	–	–	–	–	–	–	B1	–	14	
Red McClure	–	–	–	–	–	–	B	–	–	23,11	
Red Pentland Beauty	–	–	–	–	D3	–	–	A1,2	–	13,11	
Red Pontiac	–	–	A	–	C3	–	A	A1,2,C1 D1	–	23,11,21,24	
Red Rock	–	–	–	–	B2	–	–	–	–	–	
Red Skin	–	–	–	–	A1	–	A	A1,2,3,- C1	–	13,15,16,8,11	
Red Skin Russet	–	–	–	–	B2	–	–	–	–	–	

Variety									
Red Stormont Four Eighty	—	—	—	—	—	—	A1	—	—
Red Ulster Premier	—	—	—	A1,2	—	—	A5,B1	—	13,11
Red Warba	—	D	—	—	—	—	—	—	23,11
Reflecta	—	A	—	—	C	A	—	—	11,4,21
Regale	—	A	—	—	—	A	B1	—	4,11,21
Regent	—	A	—	—	—	—	B1	—	11,14
Regiment	—	—	—	—	—	—	A1	—	14
Regina	—	A	—	A1,B2	—	—	B1,4	C1	11,4,14
Reichskanzier	B1	—	—	—	—	—	—	—	3
Reina	—	A	—	A1,2	—	—	—	—	11,21
Reine Laure	—	A	—	—	—	A	—	—	4,11,21
Reliance	—	—	—	B2	—	—	—	—	11
Remedy	—	A	—	—	—	—	—	—	11
Remona	—	—	—	B1,2	—	—	—	—	11
Reneta	—	D	—	—	—	—	—	—	11
Renova	—	A	—	—	C	A	A1,C1	—	11,4,13
Renska	—	—	—	—	C	A	—	—	32,31
Rental	—	A	—	—	D	A	—	—	32
Resident	—	A,C	—	—	—	A	—	—	11,4,21

Variety	Belgium	Denmark	France	Greece	Ireland	Italy	Netherlands	United Kingdom	FR Germany	Descriptive reference	Synonym
Resonant	–	–	A	–	A1,2	–	A	–	–	11,12	
Response	–	–	–	–	–	–	–	B1	–		
Resy	–	B1	A	–	–	C	A	–	–	10,11,12,4	
Revelino	–	B1,C3	A	–	A1,2	–	A	A1,3,5, C1	A2	11,1,13,18,31 32	
Revenu	–	–	B	–	–	–	–	–	–	11	
Rheinhort	–	B1	A	–	–	–	B	–	D1	18,19,11,21	
Rhoderick Dhu	–	–	–	–	B2	–	–	B1	–	17,8	
Rhona	–	–	–	–	–	–	–	A1,2,5	–		
Richters Jubel	–	–	D	–	–	–	–	–	–		
Ridgeway Rossmore	–	–	–	–	B2	–	–	–	–		
Rika	–	–	–	–	–	D	–	–	–		
Rila	–	–	–	–	–	–	B	–	–	11	
Risa	–	–	–	–	–	–	–	–	D1	11,19	
Rival	–	–	D	–	A1,2	–	A	B1	–	11	

Variety									Ref.	Synonyms
Robijn	D	—	—	B1	—	A	—	—	11	
Robinia	D	—	—	—	—	—	—	—		
Robusta	D	B1	—	—	—	—	—	—	11	
Rocks	—	—	—	—	—	—	A1	—	8,16	
Rød Ankergaard	—	B1	—	—	—	—	—	—	2	Rod Ankergaard
Rode Eerstelingen	D	—	—	—	—	A	A1,C1	—	12,5	Rode Ersteling
Rode Ersteling	*See Rode Eerstelingen*								10,11,21	Rode Eerst-elingen
Rode Industrie	D	—	—	—	—	—	—	—		
Rode Pipo	A	B1,C3	—	—	C	A	A1,3,C1	—	11,12,21	
Rode Star	D	B1,C3	—	A2,B1	—	A	A1	—	11	
Rode Star Famboise	D	—	—	—	—	—	—	—		
Rohlicky	A	—	—	—	—	—	—	—		
Roland II	D	B1	—	—	—	—	—	—	14	
Romano	—	—	—	A2,B1, C3	C	A	A1,3,5, C1,4	—	10,11,12,13	
Romano (A)	—	—	—	C3	—	—	—	—		
Romanze	—	—	—	—	—	—	A1	—	18,11,21	
Romeo	A	—	—	—	—	—	—	—		
Romula	—	—	—	—	—	A	—	—	32,31	

Variety	Belgium	Denmark	France	Greece	Ireland	Italy	Netherlands	United Kingdom	FR Germany	Descriptive reference	Synonym
Ronda	–	–	–	–	–	–	A	–	–	11,21	
Ronde Jaune du Tregor	–	–	D	–	–	–	–	–	–	5	
Ronea	–	–	–	–	–	–	–	–	A1	18,11,21	
Ropta	–	–	–	–	–	–	–	–	D1	11	
Ropta F292	–	–	–	–	–	D	–	–	–		
Rosa	–	–	A,C	–	–	–	A	C4	D1	4,11,19,21	
Rosa (USA)	–	–	–	–	–	–	–	A2	–		
Rosabelle	–	–	A	–	–	C	–	–	–	4,11,21	
Rosafolia	–	–	A	–	–	–	–	–	–	29	
Rosalie	–	–	A,C	–	–	–	A	–	–	4,11,21	
Rosamunda	–	–	–	–	–	–	–	B1	–	11,21	
Rosanna	–	–	A	–	–	–	A	–	–	11,21	
Rosabelle	–	–	–	–	–	C	–	–	–		
Rosa Violette	–	–	D	–	–	–	–	–	–		

Variety										
Rose de Cherbourg	–	–	D	–	–	–	–	–	–	5
Rosedor	–	–	A	–	–	–	A	–	–	11
Roseval	–	–	A,C	–	B1	–	B	A1	–	4,11,21
Rosine	–	–	A	–	–	–	A	–	–	4,11,21
Rosita	–	–	A	–	–	–	–	A2	–	30
Roslin Castle	–	–	–	–	–	–	–	A1,2,3	A2,3	11,13
Roslin Chania	–	–	–	–	–	–	–	A1	–	
Roslin Eburu	–	–	A	–	B1	–	–	A1,2	–	13
Roslin Elemteita	–	–	–	–	–	–	–	A1	–	,
Roslin Mt Kenya	–	–	–	–	–	–	–	B1	–	
Roslin Riviera	–	–	–	–	B2	–	–	A1,2	–	13
Roslin Sasumua	–	–	–	–	–	–	–	A1,2	–	14
Rostovskij	–	–	D	–	–	–	–	–	–	
Rosva	–	B1,C3	–	–	–	–	A	–	D1	1,11,21
Rote Niere	–	–	D	–	–	–	–	–	–	
Rotweissragis	–	–	D	–	–	–	–	–	–	
Rougeor	A,C	–	–	–	–	–	–	–	–	11,21
Roxane	A,C	–	–	–	–	–	–	–	–	11,21
Roxy	–	–	–	–	–	–	–	–	A2	18,11,21

Variety	Belgium	Denmark	France	Greece	Ireland	Italy	Netherlands	United Kingdom	FR Germany	Descriptive reference	Synonym
Royal Kidney	–	–	D	–	A2	–	B	A1,2,3, C1	–	13,15,16,11	Queen Mary
Rua	–	–	–	–	–	–	–	A1	–	11,14	
Rubin	–	–	D	–	–	–	–	–	–	11	
Rubingold	–	–	A	–	–	–	–	–	D1	28	
Rubinia	–	–	–	–	–	–	A	A3,B1, C1	–	32	
Ruby Queen	–	–	–	–	A2,B2	–	–	–	–	8,21	
Rubynia	–	–	–	–	B2	–	–	–	–		
Runo	–	–	A	–	–	–	–	–	D1	11	
Rural New Yorker	–	–	–	–	–	–	–	A1	–	23,17	
Russet Burbank	–	B1	–	–	B2	–	A	A1,C1	–	11,13,23	
Russet Burbank (Seedling)	–	–	–	–	B2	–	–	–	–		
Russet Conference	–	–	–	–	–	–	–	A1,3	–	13,11,21	
Rutt	–	–	–	–	–	C	–	–	–		
Ryecroft Purple	–	–	–	–	–	–	–	A1	–	16,17	

Variety											
Sabina	—	—	D	—	—	—	—	B1	—	11,21	
Sable	—	—	—	—	—	C	B	—	—	11,24	
Sabonete	—	—	—	—	—	—	A	—	—	11	
Sack Filler	—	—	—	—	B2	—	—	—	—	8	
Saco	—	B1	—	—	A2,B1	—	A	A1,B4,C4	—	23,11,14	
Saga	—	—	—	—	—	—	B	—	—	11	
Sagitta 11	—	B1	—	—	—	—	—	—	—	11	
Sahel	—	—	A	B	—	—	A	—	—	4,11,21	
Saida	—	—	A	—	—	—	A	—	—	4,11,21	
Saito	—	—	—	—	—	—	—	—	D1	—	
Salinka	—	—	—	—	—	C	—	B1	—	11,21	
Salvia	—	—	A	—	—	—	—	—	—	11	
Sandra	A	—	—	—	—	—	A	—	—	31,32	
San Michele	—	—	—	—	—	—	B	—	—	11,28	San Michele (1)
San Michele (1)	—	—	C	—	—	—	—	—	—	—	San Michele
Sante	—	—	A	—	—	C	A	A1,3,C1	—	12,31	
Saphir	—	—	A	—	—	—	A	B1	A2	18,19,11,21	
Sapora	—	—	A	—	—	—	—	—	—	11	
Saskia	—	B1,C3	A	—	A1	—	A	B1	A2	10,11,12,1,4	

Variety	Belgium	Denmark	France	Greece	Ireland	Italy	Netherlands	United Kingdom	FR Germany	Descriptive reference	Synonym
Satapa	–	–	D	–	–	–	–	–	–	23	Waseca
Satelliet	–	–	–	–	–	–	A	–	–	31,32	–
Satisfaction	–	–	–	–	–	–	–	B1	–	14	
Saturna	–	B1,C3	A	–	A1,2	–	A	A2,B1	B2,D1	10,11,12,1,4,31	
Sarlotte	–	–	–	B	–	–	–	–	–		
Saucisse	–	–	B,D	–	–	–	B	–	–	11	Saucisse Rouge
Saucisse Rouge	*See* Saucisse										Saucisse
Sava	–	B1,2,C3	–	–	–	–	A	C1	–	1,11,32	
Scaldia	A	–	–	–	–	–	–	–	–	11,21	
Scala	–	–	–	B	–	–	–	B1,C1	–		
Schenklander	–	–	A	–	–	–	–	–	–	11,32	
Schlesien	–	–	D	–	–	–	–	–	–	28	
Schoolmaster	–	–	–	–	–	–	–	B1	–		
Schoolmeester	–	–	–	–	–	–	B	–	–	11	

Variety											Synonym
Schwalbe	–	B1	–	–	–	–	A	–	–	11	
Scotston Supreme	–	–	–	–	–	–	–	B1	–	–	
Sebago	–	B1	D	B	B2	–	A	B1	D1	23,11,14,21	
Secunda	–	–	–	–	–	–	–	–	D1	11	
Sedina	–	–	–	–	–	–	–	–	A2	18,11,21	
Sefton Wonder	–	–	–	–	–	–	–	B1	–	17,8	
Selma	–	–	C	–	–	–	–	–	A2	18,11,21	
Semenic	–	–	–	–	–	C	–	–	–	–	
Senator	–	B1	–	–	–	–	A	–	–	12,31	
Semena	–	–	–	–	–	–	–	B1,C1	–	–	
Senta	–	–	–	–	–	–	–	–	D1	11,21	
Sequoia	–	–	D	–	–	–	–	B1	–	23,11	
Serana	–	–	–	–	–	–	A	–	–	32	
Serrana	–	–	–	–	–	–	–	A2,C4	–	–	
Sevenster 70–7–3=EKE	–	–	–	–	–	–	A	–	–	31,32	Eke
Severianin	–	–	D	–	–	–	–	–	–	27,11	Sewerjanin
Severnaja Rose	–	–	D	–	–	–	–	–	–	11	
Shamrock	–	–	A	–	A1,2	–	–	B1,2	–	8,11,16	
Sharpes Express	–	B1	D	–	A1,2,C3	–	–	A1,2,3,5 C1	–	13,15,16,17, 7,8,11,21	Sydens Dronning

Variety	Belgium	Denmark	France	Greece	Ireland	Italy	Netherlands	United Kingdom	FR Germany	Descriptive reference	Synonym
Sharpes Pink Seedling	–	–	–	–	–	–	–	B1	–	16	
Sharpes Victor	–	–	D	–	B2	–	–	A1	–	16,17,8	
Sheriff	–	–	–	–	A1,2	–	–	A1,2,3,5	–	13,11,21	
Shepody	–	–	–	B	–	–	–	C1	–		
Shetland	–	–	–	–	–	–	–	B1	–		
Sheena	–	–	–	–	–	–	–	A1,2,C1	–		
Shurchip	–	–	–	–	–	–	B	–	–	11	
Sibirjak	–	–	D	–	–	–	–	–	–	11	
Sickingen	–	–	D	–	–	–	–	–	–	5	
Sieglinde	–	B1,C3	A	–	B2	C	A	A1	B2,D1	18,19,1,4,11	Sieglineol
Sientje	–	B1,C3	A	–	A2,B1	–	A	A1,B4	–	1,4,7,13	
Sierra	–	–	–	–	–	D	–	–	–		
Sierra Volcan	–	–	C	–	–	–	–	–	–	11	
Sigma	–	–	–	–	–	–	–	–	A1,B1	11,18,21	
Simone	–	–	–	–	–	–	A	–	–	32	

Variety										
Simson	A	—	—	—	—	·A	—	—	11	
Sinaeda	—	A	—	—	—	A	B1	—	11,21,31,32	
Sirco	—	A	—	—	D	A	B1	D1	10,11,4	
Sir John Llevelyn	—	D	—	—	—	—	—	—	17	Eclipse
Siro	—	—	—	—	—	A	—	—	32	
Sirtema	B1,C3	A,C	A1,2	C	A	A	A1,3,B4	—	10,11,12,1,4	
Sitta	—	—	—	—	—	—	—	D1	11	
Sjitskii	*See* Losicki									Losicki
Skerry Blue	—	—	B2	—	—	—	A1	—	8	
Skerry Champion	—	—	B2	—	—	—	—	—	7,11,21	
Skorospelka No. 1	—	—	—	—	B	—	—	—	11	
Skirza	—	—	—	—	—	—	C1	—		
Skurvres fort Fra Svalof	B1	—	—	—	—	—	—	—		
Skutella	B1	A	—	—	—	—	—	—	11	
Smaragd	—	A	—	—	—	—	—	D1	19,11	
Snowchip	B1	—	—	—	—	—	—	—		
Snowdrop	—	—	B2	—	—	—	—	—	17,8	
Snowdrop (Bolter)	—	—	B2	—	—	—	—	—		Witch Hill Early, Witch-hill

Variety	Belgium	Denmark	France	Greece	Ireland	Italy	Netherlands	United Kingdom	FR Germany	Descriptive reference	Synonym
Snowflake	–	–	–	–	B2	–	–	–	–	8,11	
Sokol	–	–	–	–	–	–	B	–	–	20,11,21	
Solanum	–	–	A	–	–	–	B	–	–	11	
Solene	–	–	–	–	–	D	–	–	–		
Solist	–	–	–	–	–	–	A	–	–	31,32	
Sommerkrone	–	–	C	–	–	–	–	–	D1	11,19	
Sommerniere	–	–	–	–	–	–	–	–	D1	11,19	
Sommerstarke	–	–	D	–	–	–	A	–	A2	18,19,11,21	
Southesk	–	–	–	–	B2	–	–	B1,2	–	8,11	East Nevk
Souvenir	–	–	D	–	–	–	A	–	–	11	
Sovietskij	–	–	D	–	–	–	–	–	–	11,27	
Spartaan	–	–	A	–	B1,2	–	A	A1	–	11,12,4	
Spatrot	–	–	D	–	–	–	–	–	D1	28	
Spatz	–	–	A	–	–	–	–	–	–	11	
Spekula	–	B1	–	–	–	–	–	–	–	11	

Variety										
Spiertnicks	–	–	–	–	–	–	–	–		
Splendor	–	–	D	–	–	–	A	–	11,14	
Sprys Abundance	–	–	–	–	B2	–	–	A1	8	
Spunta	–	B1,C3	A,C	B	B1	C	A	A1,2,3,5,-B4,C1,4	10,11,12,1,4	
St. Aidan	–	–	–	–	–	–	–	B1	14	
Stabiel	–	–	–	–	–	–	A	–	31,32	
St. Malo Kidney	–	–	–	–	–	–	A	A1		Colossal
Stania	–	–	A	–	–	–	A	–	11,4,31	
Starkeragis	–	B1	D	–	–	–	–	–	3	
Starkereiche	–	B1	–	–	–	–	–	–	3	
Steffi	–	B1	A	B	–	–	–	A2	18,1,11,21	
Stella	–	–	A	–	–	–	A	D1	4,11,21	
Stemester	–	–	–	–	–	–	–	B1,C1		
Stina	–	–	–	–	–	–	–	B1	11,21	
Stirling Castle	–	–	–	–	–	–	–	A1	8,16	
Stormont Dawn	–	–	D	–	B2	–	–	A1	16,11	
Stormont Enterprise	–	–	–	–	A2,B1	–	–	A1,2,3,5	13,15,11,21	
Stormont Star	–	–	–	–	–	–	–	B1		
Stormont 4-80	–	–	–	–	B1,2	–	–	A1	13,16,11,21	

Variety	Belgium	Denmark	France	Greece	Ireland	Italy	Netherlands	United Kingdom	FR Germany	Descriptive reference	Synonym
Strath	–	–	–	–	–	–	–	A1,2	–	11,14	
Striped Champion	–	–	–	–	B2	–	–	–	–	14	
Subliem	–	–	–	–	–	–	A	–	–	32	
Sucevita	–	–	–	–	–	C	–	–	–		
Suevia	–	–	D	–	–	–	–	–	–	11	
Summit	–	–	–	–	–	–	–	B1	–	16,8	
Sunia	–	–	–	–	–	–	B	–	D1	19,11	
Super	–	–	–	–	–	C	–	–	–		
Superior	–	–	A	–	–	C	A	–	–	11,24	
Suprise	–	–	A	–	–	–	A	–	–	11,12,17,21	
Susanna	–	–	D	–	–	–	–	–	–	11	
Suttons Abundance	–	–	–	–	B2	–	–	A1	–	16	
Suttons Angus Beauty	–	–	–	–	–	–	–	A1	–	14,16	
Suttons Angus Gem	–	–	–	–	–	–	–	A1	–	16,11	
Suttons Commander	–	–	–	–	–	–	–	B1	–	14	

Sharpes Express

Suttons Early Regent	–	–	–	–	–	–	–	B1	16
Suttons Foremost	–	–	–	A1,2	–	–	A1,2,3, B4	–	13,15,16,11
Sutton Olympic	–	–	–	–	–	–	A1	–	14,16
Suttons Preference	–	–	–	–	–	–	A1	–	14,16
Suttons Victoria	–	–	–	B2	–	–	–	–	
Svalof Birgitta	–	B1	–	–	–	–	–	–	3
Sviazskij	D	–	–	–	–	–	–	–	
Sydens Dronning	*See* Sharpes Express								
Taborky	D	–	–	–	–	–	–	–	11
Tachibana	–	–	–	–	–	B	–	–	11
Tahi	–	–	–	–	–	–	A1	–	11,14
Taiga	–	B1	–	–	–	A	B1	A2	18,11,21
Talovskii 110	–	–	–	–	–	B	–	–	11
Tamara	–	–	–	–	–	–	–	D1	19,11
Tammiston Aikenen	D	–	–	–	–	–	–	–	27
Tanja	A	–	–	B1	–	A	–	–	11,21
Tarpan	–	–	–	–	D	–	–	–	
Tasso	A	–	–	–	–	A	–	A2	18,19,4,11

Variety	Belgium	Denmark	France	Greece	Ireland	Italy	Netherlands	United Kingdom	FR Germany	Descriptive reference	Synonym
Tawa	–	–	–	–	B1	–	–	–	–	23,11	
Tayside	–	–	–	–	–	–	–	B1	–	16	
Tecka	–	B1	–	–	B2	–	–	–	–		
Tekla	–	–	–	–	–	–	–	–	–	2	
Tella	–	–	–	–	–	–	–	B1	–		
Temp	–	–	–	–	–	–	A	–	–	11	
Templar	–	–	–	–	B2	–	–	B1	–	16,17,8	
Tempora	–	–	A	–	–	–	–	–	–	18,11,21	
Tenor	–	–	–	–	–	C	–	–	–		
Terrina	–	–	–	–	–	–	–	–	A2,D1	18,11,21	
Tertus	–	B1,C3	–	–	–	–	A	–	D1	1,11,21	
Teton	–	B1	D	–	–	–	–	–	–	23,11,24	
Thalassa	–	–	–	–	–	C	–	–	–		
Thalia	–	–	–	–	–	–	–	–	D1	19,11	
The Alness	–	–	–	–	–	–	–	B1	–	16	

Variety											
The Baron	–	–	–	–	B2	–	–	A1	–	16,8	
Theresa	–	–	–	–	–	–	A	–	–	31,32	
The Towse	–	–	–	–	–	–	–	B1	–	8	
Thola	–	–	B	–	–	–	A	–	–	11,31	
Thomana	–	–	A	–	–	–	A	B1	A2	18,11,21	
Thome Black	–	–	–	–	B2	–	–	–	–	8	
Thomes	–	–	–	–	B2	–	–	A1	–	8	
Thorbecke	–	–	–	–	–	–	B	–	–	11	
Thorma	–	–	D	–	–	–	–	–	–	11	
Thynia	–	–	A	–	A1,2	–	A	B1,4	–	11,14,21	
Thyra	–	–	A	–	–	–	–	–	D1	19,11	
Tidlig Rosen	–	B1,C3	–	–	–	–	–	–	–	11,21	Early Rose
Tiger	–	–	–	–	–	–	–	–	D1		
Timate	–	–	–	–	–	–	A	–	–	32	
Tinwald	–	–	–	–	B2	–	–	–	–		
Tinwald Perfection	–	–	–	–	–	–	–	B1	–	16,17,8	
Titana	–	–	–	–	–	–	–	A3,B1	A2,D1	18,11,21	
Titia	–	–	–	–	–	–	A	–	–	32	
Tiva	–	–	–	–	–	–	–	–	–	2	
Titania	–	–	–	–	–	–	–	A3,B1	–		

Variety	Belgium	Denmark	France	Greece	Ireland	Italy	Netherlands	United Kingdom	FR Germany	Descriptive reference	Synonym
Tomasa Condemayta	–	–	C	–	–	–	–	–	–		
Tombola	–	–	A	–	A1,2	–	A	–	–	11,14	
Tondra	–	–	–	–	–	–	A	B1	D1	19,11	
Tonika	–	–	–	–	–	–	–	–	A2	18,11,21	
Topi	–	–	D	–	–	–	–	–	D1	11	
Torero	–	–	D	–	–	–	–	–	–	11	
Total	–	–	A	–	–	–	–	–	–	11	
Trevor	–	–	A	–	–	–	–	–	–	11	
Triumf	–	–	–	–	–	–	B	–	–	11	
Triumph	–	–	D	–	–	–	–	–	–	23,11	Red Bliss
Triumph de Kerkov	–	–	D	–	–	–	–	–	–		
Triumph = Red Bliss	–	–	–	–	–	–	B	–	–	11	
Trophee	–	–	A	–	A1,2	–	–	–	–	11	Triumph
Troubadour	–	–	–	–	–	C	A	–	–	32	
Tuinparel	–	–	–	–	–	–	B	–	–	11	

Variety										
Tunika	–	B1	–	–	–	–	B	–	–	11
Turkis	–	–	–	–	–	–	–	–	B1	11
Tuskar	–	–	A	–	A1,2,C3	–	–	A1,3	–	9,11,13,21
Tylstrupodin	–	B1	–	–	–	–	–	–	–	11
Tylva	–	B1,2,C3	–	–	–	–	A	–	D1	1,11,21
Uhtomskij	–	–	D	–	–	–	–	–	–	
Ukama	–	B1,C3	A	–	A2,C3	C	A	A1,3,C1	A2	10,11,12,1,4,31
Uljanovski	–	–	D	–	–	–	–	–	–	11
Ulla	–	B1	A	–	–	–	B	–	A2	18,11,21
Ulster Beacon	–	–	–	–	A1,2	–	–	A1,B4,C4	–	14,16,11
Ulster Brevet	–	–	–	–	–	–	–	A1,B4	–	14,11
Ulster Chieftain	–	–	A	–	A2,B1	–	A	A1,2,3,5,B4,C1	–	13,15,16,7
Ulster Classic	–	–	–	–	A1,2	–	–	A1,5	–	13,11,21
Ulster Commerce	–	–	–	–	A1,2	–	–	B1	–	14,7,11
Ulster Concord	–	–	–	–	A1,2	–	–	A1,2	–	14,7,11
Ulster Cromlech	–	–	D	–	–	–	–	B1	–	14,16,11
Ulster Dale	–	–	–	–	A2,B1	–	A	A1,3,5,B4	–	- 13,16,11,21

Variety	Belgium	Denmark	France	Greece	Ireland	Italy	Netherlands	United Kingdom	FR Germany	Descriptive reference	Synonym
Ulster Earl	–	–	–	–	A1,2	–	–	B1	–	14,16,11	
Ulster Emblem	–	–	D	–	A1	–	–	A1	–	13,16,11,21	
Ulster Ensign	–	–	D	–	–	–	–	B1	–	14,16	
Ulster Glade	–	–	–	–	A1,2	–	A	A1	–	13,11,21	
Ulster Glen	–	–	–	–	–	–	–	A1	–	14,16	
Ulster Gozo	–	–	–	–	–	–	–	B1	–	14,16	
Ulster Grove	–	–	–	–	–	–	–	A1	–	14,16	
Ulster Knight	–	–	–	–	–	–	–	B1	–	14,16,11	
Ulster Lancer	–	–	A	–	–	–	–	A1,2	–	13,11	
Ulster Leader	–	–	–	–	B1,2	–	–	B1	–	14,16,11	
Ulster Magnet	–	–	–	–	B1,2	–	–	A1	–	14,16	
Ulster Malta	–	–	–	–	B2	–	–	–	–	14,16,11	
Ulster Monarch	–	–	–	–	B1	–	–	B1	–	14,8,11	
Ulster Premier	–	–	D	–	A2,B1	–	B	A1,3,5, B4	–	13,15,16,7,11	

Variety								General
Ulster Prince	–	–	D	–	A1,2	B	A1,2,3,5,– B4.C1	–	13,15,16,7,11
Ulster Ranger	–	–	–	–	A1,B2	–	A1,B4, C1	–	13,16,11,21
Ulster Sceptre	–	B1	A	–	A1,2	A	A1,2,3,5 B4,C1	–	13,15,7,21,11
Ulster Sovereign	–	–	–	–	B1,2	–	A1	–	14
Ulster Supreme	–	–	A	–	A2,B1	–	A1,B4	–	14,16,11
Ulster Tarn	–	–	–	–	A1,2	–	A1,B4	–	14,16,11
Ulster Torch	–	–	–	–	A1,2	–	A1,3	–	13,16,11,21
Ulster Viscount	–	–	–	–	B1,2	–	A1,B4	–	14
Ultimus	–	B1	A	–	A2,B1	A	B1,4,B4	B1	11,4,21
Umbra	–	–	A	–	–	–	–	–	11
Univers	–	–	–	–	–	A	B1	–	32
Universal	–	–	–	–	B2	B	–	–	11
Univita	–	B1	A	–	–	–	–	A2	18,11,21
Up to Date	–	B1,C3	D	–	A1,2,C3	B	A1,2,3,5 B4,C1	–	13,15,16,11
Uran	–	B1	–	–	–	A	–	–	20,11,21
Urgenta	A	B1,C3	A,C	–	A1	A	–	D1	11,1,4,21,22
Ursula	–	–	A	–	–	A	–	–	11
Utility	–	–	–	–	B2	–	A1	–	17,8

Variety	Belgium	Denmark	France	Greece	Ireland	Italy	Netherlands	United Kingdom	FR Germany	Descriptive reference	Synonym
Van Gogh	–	–	–	–	–	D	–	–	–		
Vakon	–	–	–	–	–	–	A	A1,C1	–	32	
Valdor	–	–	A	–	–	–	–	–	–	11	
Vale	–	–	D	–	–	–	–	–	–		
Valeria	–	–	–	–	–	–	–	–	D1	11	
Vally	–	–	A	–	–	–	–	–	A1,2	18,11,21	
Vanessa	–	–	A	–	A2	–	A	A1,3,5, C1	–	11,13,15,21	
Vanguard	–	–	–	–	–	–	–	A1	–		
Vantage	–	–	D	–	–	–	–	A1,3,5	–		
Varmas	–	–	–	–	–	–	B	–	–	11	
Vebaca	–	–	A	–	–	–	A	–	–	31,32	
Veenster	–	–	–	–	–	–	A	–	D1	11,31	
Vekaro	–	–	–	–	–	C	A	–	–	10,11,12,21	
Veloka	–	–	–	B	–	C	A	–	–	31,32	

Variety											
Venus	–	–	–	–	–	–	–	–	B1	14	
Vera	–	–	D	–	B2	–	–	–	–	11	
Verano	–	–	A	–	–	–	–	–	–	–	
Verena	–	–	–	–	–	–	A	A1,3	D1	11,13,21	Brenta
Veritas	–	–	–	–	–	–	A	B1	–	32	
Vertifolia	–	–	D	–	–	–	–	–	–	11	
Vestar	–	–	A	–	–	–	A	–	–	11,21	
Vevi	–	–	–	–	–	–	–	–	D1	11	
Victor	–	–	–	–	–	–	A	–	–	11,21	
Victoria	–	–	–	–	–	C	–	–	–	–	
Vindika	–	–	A	–	–	C	A	–	–	10,11,12,4	
Vineta	–	–	D	–	–	–	–	–	–	11	
Viola	–	–	A	–	–	–	A	–	D1	4,11,21	
Virgil	–	B1	–	–	–	–	–	–	–	–	
Virginia	–	–	–	–	–	–	–	–	D1	11,19	
Vitelotte Rouge	–	–	D	–	–	–	–	–	–	5	
Vittorini	A	–	A	B	–	C	A	–	–	11,21	
Vivaks	–	–	A	–	–	C	A	B1	–	10,11,12,4	
Vokal	–	–	A	–	–	C	A	–	–	10,11,4	
Volhovskij	–	–	D	–	–	–	–	–	–	–	

Variety	Belgium	Denmark	France	Greece	Ireland	Italy	Netherlands	United Kingdom	FR Germany	Descriptive reference	Synonym
Voran	-	-	A	-	A1,2	-	A	-	D1	19,4,7,11,21	
Vulkano	-	-	A	-	-	C	A	A1,C1	-	32	
Waalster	-	-	A	-	-	-	-	-	-	11	
Wachtel	-	-	A	-	-	-	A	-	A2	18,11,21	
Wanda	-	-	A	-	B1	-	-	-	-	19,11	
Warba	-	-	D	-	-	-	A	-	-	23,11,24	
Warta	-	-	-	-	-	-	-	-	D1	11	
Waseca?	*See* Satapa										Satapa
Wauseon	-	B1	A,D	-	-	-	A	A2	-	11,24	
Waverley	-	-	-	-	-	-	-	B1	-	8,16	
Webbs Pride	-	-	-	-	-	-	-	A1	-		
Webbs Tidling	-	B1	-	-	-	-	-	-	-	3	
Wega	-	-	A	-	-	-	-	-	A2	18,11,21	
Weisses Rassl	-	-	D	-	-	-	-	-	-	6	

Variety											
Wekaragis	—	B1	—	—	—	—	—	—	D1	19	—
Welcome	—	—	—	—	—	—	A	—	—	31,32	—
Welsa	—	—	—	—	—	C	—	B1	—	11,21	—
Werta	—	—	A	—	—	—	—	—	—	11,21	—
Westbrabanda	—	—	D	—	—	—	—	—	—	6	—
White Beauty of Hebron	*See Puritan*										Puritan
White City	—	—	—	—	—	—	—	B1	—	17,8	—
White Flowering Skerry	—	—	—	—	B2	—	—	—	—	—	—
White Fortyfold	—	—	—	—	—	—	—	B1	—	—	—
White Rock	—	—	—	—	B2	—	—	—	—	—	—
White Rose	—	—	D	—	—	C	B	—	—	23,11,24	—
Wichip	—	B1	—	—	—	—	—	—	—	—	—
Wigro	—	—	—	—	—	—	A	—	—	11	—
Wild Champion	—	—	—	—	B2	—	—	—	—	—	—
Wilja	A	—	—	—	A1,2,C3	C	Â	A1,2,3 5,C1	A2,D1	10,11,12,13	Jose
Wilja A	—	—	—	—	C3	—	—	—	—	—	—
Wilpo	—	—	D	—	—	—	—	—	—	11	—
Winda	—	—	—	—	—	—	A	—	—	31,32	—

Variety	Belgium	Denmark	France	Greece	Ireland	Italy	Netherlands	United Kingdom	FR Germany	Descriptive reference	Synonym
Wis	–	–	–	–	–	–	A	–	–	20,11	
Wisla	–	–	A	–	–	–	–	–	–	20,11	
Witchhill	–	–	–	–	–	–	–	A1	–	16	Snowdrop
Witch Hill Early	*See Snowdrop Witchhill*										Snowdrop
Witte Desiree	–	–	–	–	–	–	A	–	–	11	
Witte Ultimus	–	–	A	–	–	–	–	–	–	11	
Wonderful	–	–	–	–	–	–	–	B1	–		
Woudster	–	B1	–	–	A1,2	–	A	A1,B4	–	11,12,14,21	
Wyszoborskie	–	–	A	–	–	–	–	–	–	27	
Xenia	–	–	–	–	–	–	–	–	D1	11,21	
Yam	–	–	–	–	B2	–	–	A1	–	8,16	
Yampa	–	–	–	–	B1	–	A	–	–	23,11	
Yankee Baby	–	–	–	–	B2	–	–	–	–	8	

Yogeva Kollane	–	–	–	–	A	–	–	–	–	–
Ysselster	IJsselster	11,21	–	–	A	–	–	–	A	–
Yungay	–	30	–	–	–	–	–	–	C	–
Yulon Gold	–	–	–	–	–	–	–	B	–	–
Zazerskii	–	11	–	–	–	–	–	–	D	–
Zeeuwse Blauwe	–	11	–	–	B	–	–	–	–	–
Zeisig	–	11	–	–	A	–	–	–	D	–
Zenith	–	11,21	D1	B1	–	–	–	–	–	–
Zeeburger	–	11	–	–	–	–	B1	–	–	–

Blight R. gene differentials

R. gene status	Belgium	Denmark	France	Greece	Ireland	Italy	Netherlands	United Kingdom	FR Germany	Descriptive reference	Synonym
r	–	B1	–	–	A1	–	–	–	–		
R1	–	B1	B,D	–	A1	–	–	–	–		
R2	–	B1	B,D	–	A1	–	–	A2	–		
R3	–	B1	B,D	–	A1	–	–	–	–		
R4	–	B1	B,D	–	–	–	–	A2	–		
R5	–	B1	–	–	A1	–	A	A2	–		
R7	–	B1	–	–	A1	–	A	A2	–		
R8	–	B1	–	–	A1	–	A	A2	–		
R9	–	B1	–	–	–	–	–	A2	–		
R10	–	B1	–	–	A1	–	A	A2	–		
R11	–	B1	–	–	A1	–	A	A2	–		
R1,R2	–	B1	–	–	A1	–	–	A2	–		
R1,R3	–	B1	–	–	A1	–	–	A2	–		
R1,R4	–	B1	–	–	A1	–	–	A2	–		
R2,R3	–	B1	–	–	A1	–	–	A2	–		

R2,R4	—	B1	—	—	A1	—	—	A2	—
R3,R4	—	B1	—	—	A1	—	—	A2	—
R1,R2,R3	—	B1	—	—	A1	—	—	A2	—
R1,R2,R4	—	B1	—	—	—	—	—	A2	—
R1,R3,R4	—	B1	—	—	A1	—	—	A2	—
R2,R3,R4	—	B1	—	—	A1	—	—	A2	—
R1,R2,R3,R4	—	B1	—	—	A1	—	—	A2	—

Potato cyst nematode differential set

Seedling no. and resistance	Belgium	Denmark	France	Greece	Ireland	Italy	Netherlands	United Kingdom	FR Germany	Descriptive reference	Synonym
D22/1 (Pa1)	–	–	–	–	–	–	–	A2	–		
D23/4 (Pa2)	–	–	–	–	–	–	–	A2	–		
D31/11 (Pa1.Ro1)	–	–	–	–	–	–	–	A2	–		
D40/5 (Pa1.Pa2.Pa3 and New Leake)	–	–	–	–	–	–	–	A2	–		
D40/8 (Pa1. Pa2. Pa3.Ro1 and New Leake)	–	B1	–	–	–	–	–	A2	–	·	
D42/8 (Pa1.Pa2.Pa3 and New Leake)	–	–	–	–	–	–	–	A2	–		
D47/11 (Pa1.Pa2.Pa3. Ro1.Ro2.Ro3)	–	B1	–	–	B1	–	–	A2	–		
D49/1 (Pa1.Pa2. Pa3. Ro1)	–	–	–	–	–	–	–	A2	–		
Corsair (Pa1.Pa2.Pa3. Ro1)	–	B1	–	–	–	–	–	A2	–		
62–33–3 (Ro1.2.3.4. Pa1.2)	–	B1	–	–	–	–	–	A2	–		

58 1624–4 (Ro1.2.3)	–	B1	–	–	–	–	–	A2	–
D23/3 (Pa1)	–	–	–	–	–	–	–	A2	–
D23/4 (Pa1)	–	–	–	–	–	–	–	A2	–
65 346 19 (Ro1.2.3. 4.5)	–	B1	–	–	–	–	–	A2	–
S. kurtzianum 60.21.19 (Ro1,2)	–	B1	–	–	–	–	–	–	–
S. multidissectum P.55/7 (Pa1)	–	B1	–	–	–	–	–	–	–

Milton Keynes UK
Ingram Content Group UK Ltd.
UKHW020820141024
449569UK00008B/497

9 789061 916338